T0325839

Power Systems

Yong Li · Dechang Yang · Fang Liu
Yijia Cao · Christian Rehtanz

Interconnected Power Systems

Wide-Area Dynamic Monitoring and Control Applications

 Springer

Yong Li
Hunan University
Changsha
China

Yijia Cao
Hunan University
Changsha
China

Dechang Yang
China Agricultural University
Beijing
China

Christian Rehtanz
TU Dortmund
Dortmund, Nordrhein-Westfalen
Germany

Fang Liu
Central South University
Changsha
China

ISSN 1612-1287   ISSN 1860-4676   (electronic)
Power Systems
ISBN 978-3-662-48625-2   ISBN 978-3-662-48627-6   (eBook)
DOI 10.1007/978-3-662-48627-6

Library of Congress Control Number: 2015958342

Printed on acid-free paper

This Springer imprint is published by SpringerNature
The registered company is Springer-Verlag GmbH Berlin Heidelberg

Foreword

Interconnection has traditionally represented the essence of modern power transmission systems. However, with the introduction of faster devices in electric power systems and further utilization of power networks in wider domains, the issues pertaining to power stability issues have become more complex. The occurrence of low frequency oscillations (LFO) can affect stable and efficient operations of such interconnected power systems. However, the use of wide-area measurement systems (WAMS) in transmission systems can provide more opportunities for dynamic monitoring and control of interconnected power systems. In addition, the flexible and quick control provided by power electronics-based high-voltage direct current (HVDC) and flexible AC transmission system (FACTS) devices can increase the controllability of such interconnected systems. The optimal planning and control of an integrated system which is based on WAMS, HVDC and FACTS can provide an effective wide area damping control for enhancing the stability of interconnected power systems.

This book investigates and properly discusses the use of dynamic monitoring and control of interconnected power systems. The contents of the book are divided into two parts. The first part is on the dynamic monitoring which utilizes the wide area information of large-scale power systems. Extensive discussions on the use of advanced monitoring algorithms such as Hilbert Hunang Transform (HHT), complex orthogonal decomposition (COD), and empirical mode decomposition (EMD) are discussed in part one for identifying LFO modes and revealing dynamic characteristics of interconnected power systems. The second part of the book is on wide area stability and dynamic control. The coordination between local power system stabilizers (PSS) and wide area damping controller (WADC) is presented in part two. Further, the robust coordination between multiple WADC is proposed, and the application of optimal wide area control signals is applied to WADC. The time-varying delay effects of wide area signal are analyzed, and the advanced delay-dependent robust design approaches are proposed in part two to improve the control performance of WADC. In addition, real-time simulations, algorithm designs, hardware and software implementations are emphasized for WADC.

The authors provide original algorithms, discussions, numerical simulations in various sections of the book which signify their extensive experience with the related research in electric power systems. It is my pleasure to recommend this book to my colleagues in the power engineering community and believe strongly that the manuscript presents novel ideas in dynamic power system operations and control with extensive applications for engaging researchers as well as practitioners who carry on the related studies in electric power systems.

2015   Prof. Mohammad Shahidehpour
                                                      Bodine Professor and Director
                           Robert W. Galvin Center for Electricity Innovation
                                                   Illinois Institute of Technology
                                                              Chicago, IL, USA

Preface

With the increase in size of interconnected power systems, the network structure and the operating mode become more and more complex, which inevitability causes stability problems such as low frequency oscillations (LFO). Generally, LFO represent oscillatory interaction among multi-areas of the network. It has been an important factor for destroying the system stability, reducing the transmission capacity, and limiting the interconnected capability of network.

In recent years, more and more wide-area measurement systems (WAMS) have been applied in power systems. WAMS is the advanced and combined application of synchronized phasor measurement, communication engineering, and information technology in power systems. The aim of WAMS is to realize dynamic monitoring, analysis, and control for stable and efficient operation of the global power system. At present, WAMS research mainly concerns the following two aspects: (1) the construction and application of WAMS in smart grid and (2) the stability analysis and control based on wide-area measurements.

In interconnected power systems, the high-voltage direct-current (HVDC) technology is increasingly used for the network interconnection, and flexible AC transmission systems (FACTS) are used to provide the support of network enhancement. In this book, combining WAMS technology, the flexible and quick control capability of HVDC and FACTS is developed to implement wide-area damping control (WADC) strategies for stability enhancement of interconnected power systems.

This book intends to report the new results of WAMS application in analysis and control of power systems. The book collects new research ideas and achievements such as an online identification method for low-frequency oscillations, a delay-dependent robust design method, a wide-area robust coordination strategy, a hybrid assessment and choice method for wide-area signals, free-weighting matrices method, and its application.

The first motivation for this book is to establish a systematic, multi-scale, and comprehensive approach for estimating the oscillatory parameters, approximate mode shape and energy distribution of the dominant oscillation modes of the interconnected power systems, based on the near real-time data. A systematic method is proposed to extract the oscillation mode from the ensemble measurement matrix. Combing with the rapid development of the smart transmission network and computer technique, it is mature to establish a platform for dynamic oscillation mode estimation, analysis, and control.

The second motivation for this book is to carry out the systemized research on wide-area stability analysis and control for stability enhancement of large interconnected power systems. A flexible and quick control function of HVDC and FACTS is sufficiently developed to implement wide-area damping control strategies for solution of control problems such as how to realize control coordination among multiple WADC, how to choose optimal control-input for multiple WADC, and how to suppress the delay effects of wide-area signals on the control performance of WADC.

The main research results of this book are original from the authors who carried out the related research together for almost 6 years, which is a comprehensive summary for authors' latest research results. This book is likely to be of interest to university researchers, R&D engineers, and graduate students in electrical engineering who wish to learn the core principles, methods, algorithms, and applications of WAMS.

Outlines

This book is divided into 12 chapters. Chapter 1 introduces the status quo and trends of interconnected power systems, WAMS technology and its application in the interconnected systems, and the challenges of wide-area dynamic monitoring and control. The typical stability control problems of WADC are analyzed.

Chapter 2 introduces the theoretical foundation of LFO monitoring and analysis. The LFO phenomenon is described, and two techniques are presented to analyze LFO. One is based on system model and the other is based on measured information. The differences between these techniques are explained, and the advantages and disadvantages of each technique are compared.

Chapter 3 analyzes the shortcomings of the traditional Hilbert-Huang Transform (HHT) in identifying LFO. An improved empirical mode decomposition (EMD) is proposed to address the end effects (EEs) and specific mode-mixing. The intrinsic mode function is analyzed in time and frequency, and the normalized Hiber transform (NHT) is introduced. The improved HHT, which integrates the improved EMD and NHT, is proposed to calculate the oscillatory parameters of the single measured signal.

Chapter 4 presents a relative phase calculation algorithm (RPCA) to explore the spatial distribution of the specific oscillation mode. The concepts of node contribution factor (NCF) and approximate mode shape (AMS) are proposed to describe the phase information of specific oscillation mode based on the multi-measured signals. By combining the improved HHT and RPCA, a nonlinear hybrid method

(NHM) is proposed, which can be used to not only provide the oscillatory parameters of single measured signal and NCFs of every mode, but also to calculate the AMSs of the oscillation modes.

Chapter 5 analyzes the temporal and spatial characteristics of oscillation modes in power system, by using the complex orthogonal decomposition (COD). In order to realize the near real-time application, three different COD methods, including the complex eigenvalues decomposition (C-ED), complex singular value decomposition (C-SVD), and augmented matrix decomposition (AMD), are compared under the different sizes of ensemble measurement matrixes. The measured data from wide-area-protector (WAProtector) is used to verify the effectiveness of the near real-time application of the COD-sliding window recursive algorithm (SWRA).

Chapter 6 presents an overall framework of WADC. The control concept and operating principle of WADC are investigated by studying a single-machine infinite-bus (SMIB) with shunt-type FACTS device. The system linearized modeling method (direct feedback linearization, DFL) is used for system modeling.

Chapter 7 proposes a sequential design and global optimization (SDGO) method to optimize local and wide-area controllers simultaneously. The technical concept and implementation flowchart of this method are described. The modal analysis is used for phase-compensation design of local power system stabilizers (PSS) and HVDC-WADC, and a global optimization method is presented to find suitable control gains for both PSS and HVDC-WADC. The eigenvalue analysis and nonlinear simulation on typical HVDC/AC interconnected systems are carried out to validate the proposed SDGO method.

Chapter 8 proposes a wide-area robust coordination approach for multiple HVDC- and FACTS-WADC to damp multiple inter-area oscillation modes of interconnected power systems. The architecture of wide-area control network (WACN) is presented with the advanced control ability for enhancing the overall stability. The multi-objective mixed H_2/H_∞ control synthesis is used for robust design of HVDC- and FACTS-WADC. The robustness of the closed-loop system with the designed multiple WADCs is evaluated at different operating scenarios.

Chapter 9 proposes a hybrid method to assess and select optimal input signals for multiple WADCs. An index of the relative residue ratio (RRR) is defined to pre-select input signal candidates for multiple HVDC- and FACTS-WADC. The steady-state value and frequency response curve of the elements of relative gain array (RGA) are used to determine the optimal control pairs for multiple WADCs. The proposed method has the advantage of effectively reducing or even eliminating the interaction among multiple controllers.

Chapter 10 presents a new linear design approach on the robust WADC of interconnected power systems. The free-weighting matrices are introduced to convert the optimization object with nonlinear matrix inequality constrains into a set of LMI constraints. A nonlinear optimization algorithm is presented to search the optimal control gain with the maximum delay independent of the wide-area feedback control signals.

Chapter 11 presents the hardware and software design and implementation for WADC. A hardware-in-the-loop (HIL) test system based on the RT-LAB platform®

is established. Three WADC algorithms, i.e., the phase-compensation method, the delay-dependent state-feedback method, and the delay-dependent dynamic output-feedback method are introduced. The software designs of these algorithms are carried out, and the flowcharts are proposed for implementing algorithms in the environment of hardware. A typical interconnected power system is modeled in RT-LAB®, the closed-loop test has been done to validate the proposed control concept and controller design methods in conditions with time-varying delays, and compare the damping performance of WADC using different algorithms.

Chapter 12 presents the design and implementation of parallel processing in embedded phasor data concentrator (PDC) application for monitoring and stability enhancement of interconnected power systems. The structure of an interconnected system equipped with an embedded system with PDC and FACTS-WADC applications is established. The fundamental modules of the embedded system are designed, and the embedded PDC application is implemented on the evaluation kit EVK1100 from Atmel®. The main program workflow of parallel processing in embedded PDC and WADC applications is designed and presented. The closed-loop experiment is carried out to validate the designed results.

Acknowledgments

This book project is supported by the Nature Science Foundation of China (NSFC) under Grant 51377001, 61304092, 51520105011 and 61233008, and by the International Science and Technology Cooperation Program of China under Grant 2015DFR70850.

The authors thank Prof. Tapan K. Saha and Dr. Olav Krause (University of Queensland), Prof. Ryuichi Yokoyama (Waseda University), Prof. Kwang Y. Lee (Baylor University), Prof. Mohammad Shahidehpour (Illinois Institute of Technology), Prof. Min Wu and Prof. Yong He (China University of Geosciences), Prof. Longfu Luo (Hunan University), Prof. Pin Ju (Hohai University), Dr. Ulf Häger and Dr. Kay Görner (TU Dortmund), Prof. Shijie Cheng (Huazhong University of Science and Technology), and Prof. Yuanzhang Sun (Wuhan University) for their great support and valuable comments.

Special thanks go to the postgraduate students Ms. Yang Zhou, Ms. Xia Li, and Mr. Xingyu Shi for their contribution and proofreading. Finally, each author thanks the long-term support and encouragement from their family.

2015   Yong Li
                                                                              Dechang Yang
                                                                                  Fang Lin
                                                                                  Yijia Cao
                                                                          Christian Rehtanz

Contents

Chapter 1
Introduction

In this chapter, the situation and development of electric interconnected systems are introduced first. The typical stability problems suffered by large interconnected systems are paid attention to. Later, the wide area measurement systems (WAMS) technology and its application in interconnected systems are investigated. A brief summary of the literature describing the approaches to analyze the low frequency oscillation (LFO) phenomenon from system model and measured data is provided. Finally, based on these introductions, the challenges of wide area dynamic monitoring and control are also discussed.

1.1 Status Quo and Trends of Interconnected Systems

Nowadays, the continued interconnected enhancement of regional electric networks is the developing trend of modern power systems worldwide, such as the interconnection of Europe networks (UCTE) [1, 2], the Japan power grid [3, 4], the national grids of China (CSG) [5, 6], and the North American Power Grids [7, 8]. Basically, the interconnection of electric networks can efficiently utilize various power resources distributed in different areas, and achieve the optimal allocation of energy resources. It is of advantage to optimize the economic dispatching of power energies, and get the relative cheaper power consumes, which means the decrease of system installed capacity and investment. Moreover, during fault or disturbance operating conditions, it can further provide supporting power for each area interconnected network, which can increase the reliability of the generation, transmission, and distribution systems.

However, for interconnected electric networks, the LFOs special inter-area oscillations (IAOs) are easily excited when there are faults (e.g., line-to-ground fault) or disturbances (e.g., line outage or load shedding) in the systems. Such oscillation phenomena are not typical but have a general meaning for current interconnected networks. Take some practical interconnected systems in the worldwide as examples:

© Springer-Verlag Berlin Heidelberg 2016
Y. Li et al., *Interconnected Power Systems*, Power Systems,
DOI 10.1007/978-3-662-48627-6_1

1. *Chinese power systems.* The PMU/WAMS has wide application in Chinese power systems. Totally, the Chinese power systems mainly include Central, Northern, Eastern, Southern, Northeastern, and Northwestern China's networks. More than 105 phasor measurement units (PMUs) and WAMS are distributed in these interconnected systems. The research indicates that there is a typical IAO mode with the oscillation frequency around 0.26 Hz [6]. This oscillation mode causes the power instability on the interconnected lines, and inevitably limits the interconnected ability. Thus, the necessary damping strategy should be performed to stabilize such oscillation mode.

2. *European interconnected network (UCTE).* Till date, the European power system has developed a large-scale interconnected network whose area extends from Poland to Portugal in the East–West direction and from Denmark to Greece in the South–North direction. This big network covers 28 countries synchronously interconnected with 32 transmission system operators (TSOs). In the practical operating experience, the IAOs have been detected from the real-time operating data. There are typical oscillations among four areas (Athens, Stuttgart, Seville, and Algiers). The oscillation frequency fp and damping ratio of the dominate oscillation mode are0.15 Hz and 5.3 %,respectively [2], which is the typical IAO mode.

3. *Japan Western 60 Hz electric networks.* In Japan, there is one eastern 50 Hz and one western 60 Hz electric networks, respectively. The 60 Hz system, consists of six major electric power companies [3], where each power systemis operated by each power company, and power is interconnected through 500 kV transmission lines. In practice, such interconnected systems also exist IAOs, which can be detected by the PMUs located in campuses of Nagoya Institute of Technology, Fukui University, Osaka University, Hiroshima University, Tokushima University, Kyushu Institute of Technology, and Miyazaki University [3]. The IAOs are around 0.4 Hz oscillation frequency. These curves are obtained by monitoring the phase difference between different areas. From these, it can be found that there are obvious long-term oscillations in the Western 60 Hz electric networks.

4. *The North American power system.* The North American power system is a typical large-scale interconnected system with three different electric areas [7], where the Eastern area is an important energy consuming area and consists of a lot of heavy loads. The WAMS, which bases on the PMUs, hasgot a certain application in the North American power system. Hence, it changes conveniently to monitor the dynamic behavior of the large-scale interconnected system. On August 11, 2009, some disturbance occurred in New York area, which led to a 0.20 Hz IAOs. From the mode shape of the 0.20 Hz dominant mode, it can be seen that the IAOs represent as the interaction between the Northeastern region and the Northwestern and the Southern regions of the Eastern area [7].

From the description above, it is clear that the stability of interconnected system is ahot topics in transmission network. Moreover, the LFO, especially the IAO mode, has become an important factor threatening the security of power system.

Against this background, it is necessary to set up a systematic technique which is able to assess the dynamic stability of the bulk power systems timely and efficiently.

1.2 Stability Problems of Interconnected Systems

With the interconnected power systems, it is possible to optimize the generation of various energies (e.g., thermal energy, hydroelectric energy, wind energy, etc.) distributed in different areas, to realize the global optimal scheduling for power utilization, and to optimize the power economic dispatch. Besides, it provides the capability of mutual power support from one area to another, which is good to the reliable and secure operation of large-scale power systems.

However, with the upscaling of interconnected systems, the grid structure and system operating mode become more complex, which inevitably brings some typical stability problems. According to the physical nature of system dynamic behavior, power system's stability can be mainly classified into three categories [9]: rotor-angle stability (angle stability), frequency stability, and voltage stability, as shown in Fig. 1.1.

The angle stability reflects the cooperating ability of generators interconnected by transmission lines. When the power system suffers a disturbance (large-disturbance such as line-to-ground fault, or small-disturbance such as load shedding), the mutual oscillation, between the rotors of some generators in one area against those of some generators in another area, is easily excited. If the power system lacks sufficient damping, this kind of oscillation phenomenon originated by the power generation facilities will continue and spread further to the transmission facilities, which inevitably leads to power oscillations in the tie-lines. In practice, power oscillations in the large interconnected systems usually represent LFO with

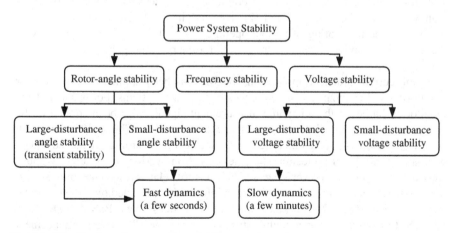

Fig. 1.1 Classification of power system stability

an oscillation frequency generally varying between 0.2 and 3.0 Hz [10]. If there is no effective oscillation damping control, the angle stability generally represents fast dynamics with the duration of a few seconds.

LFO is a typical angle stability problem existing in large interconnected systems. Especially for the IAO with an oscillation frequency of 0.2–0.8 Hz [11], it has become a key factor of endangering the operating reliability, reducing the transmitted power capacity, and limiting the interconnected ability of large electric networks. Furthermore, undamped power oscillations can excite voltage swings (voltage stability problem) of some buses distributed in different areas, and at the same time lead to frequency instability at the loads. In this dissertation, LFO and IAO are investigated, and advanced damping control strategies will be proposed to stabilize these LFO and IAO modes for the stability enhancement of large interconnected systems.

1.3   WAMS Technology and Its Application in Interconnected Systems

WAMS is the advanced and combined application of synchronized phasor measurement, communication engineering, and information technology in power systems. The aim of WAMS is to realize dynamic monitoring, analysis, and control for stable and efficient operation of the global power system. At present, the WAMS research mainly concerns the following two aspects: (1) the construction and application of WAMS in smart grid; (2) the stability analysis and control based on wide area measurements.

The structure of WAMS mainly includes the following three parts: (1) PMU devices distributed in different areas for measuring local operating variables (e.g., voltage and current); (2) A monitoring and control system located in the power system dispatch center; (3) A digital communication network in charge of information exchange between (1) and (2). By applying WAMS, it is convenient to perform online monitoring of remote operating variables, which is good to reveal the dynamic behavior of the power systems for energy management and decision-making of TSOs.

In the current practical application, WAMS has been applied successfully for monitoring and analyzing dynamic processes in power systems. For instance, LFO monitoring based on the wide-area measurement signals have been applied in the European interconnected network [12], the State Grid of China [13], the North American Grid [14], and other networks. However, regarding to the advanced wide-area control and protection strategies based on WAMS, there are few applications in the practical power systems, and most of related researches only stay at the theoretical stages. In fact, the wide-area information provided by WAMS can be further utilized by different kinds of control devices (e.g., PSS, HVDC, and FACTS) to form a wide-area control strategy for the overall stability enhancement of large power systems. As the optimal control signal can be selected on a global level and not only locally, the wide-area control can achieve more effective performance than the local control. Now, in China Southern Grid, the wide-area stability control is being considered as the supplemental control of multi-terminal HVDC transmission systems for damping IAO of the interconnected systems [15, 16].

1.4 Low Frequency Oscillation Analysis Methods

In this section, a brief summary of the literature describing the approaches to analyze the LFO phenomenon from system model and measured data is provided.

Model-based method has been widely utilized to extract the modal information as well as to identify the transfer function and to design the controller []. Generally , this method based on the system model or device parameters can be classified into two kinds: one is the linear methods, such as the eigenvalue analysis [17–20]; the other is nonlinear methods, for example, time-domain simulation and normal form analysis [21]. The key idea of the former is that the whole system is linearized around a specific operating point. Then the system model is described as the state equations. The eigenvalues, eigenvectors, and participation factors of the state matrix are calculated [17, 22]. The oscillatory parameters including the frequency and damping ratio of each mode are determined by the eigenvalues. The relationships among different modes or state variables as well as the mode shapes are described based on the eigenvectors [2217–20]. Furthermore, the sensitivities of eigenvalues are the basis for the controller design. Usually, the eigenvalue analysis methods are divided into full eigenvalue and part eigenvalue analyses. However, it is difficult to linearize the whole system after a big disturbance. Under this condition, eigenvalue method is not suitable to do dynamic analysis strictly in bulk power system. In addition, there is "dimension disaster" problem if the system magnitude is huge [17]. The latter methods are considered as the standard nonlinear methods to analyze the LFO phenomena offline. The time responses of variables are obtained by solving the dynamic differential algebraic equations of simulation model using the numerical methods [23, 24]. The advantages of these approaches are that they allow the investigation of the dynamic behavior of the system under different modes and all the characteristic patterns leading to complicated phenomena can be detected. Moreover, the nonlinear behaviors of other controllers are also taken into account simultaneously. However, there are also many limitations for the application of the time-domain simulation: (1) it is impossible to trigger all the oscillation modes because the disturbance location and style are chosen specifically; (2) it is difficult to determine the oscillation characteristics only based on the limited time-domain responses; (3) time-domain simulation model is based on the detailed parameters of all elements. It is impossible to get these parameters accurately because of the benefits of manufacturers; (4) longer simulation time and heavy computation load don't make it fulfill the requirements of on-line application.

According to the features of eigenvalue analysis and time-domain simulation, both methods should be utilized mutually and in a complementary manner [25].

With the increasing utilizations of measurement devices throughout the system, especially the construction of WAMS, the methods which are based on the measured data on-time or off-time, have been paid more attention in the past few years.

A large collection of methods has been investigated for extracting model information from measured system responses. In this dissertation, these methods are divided into two groups according to the number of disposed signals. The first kind of approach is presented to calculate the oscillatory parameters of single measured signal, such as Fast Fourier Transform (FFT) [26], Prony and its improvements [27–29], Kalman filtering [30], Hilbert− Huang Transform (HHT) [31–33], Wavelet Transform (WT) [34, 35], and so on. The second kind of method is presented to extract the dynamic oscillation characteristic based on the ensemble measurement matrix, such as, multi-Prony technique [36, 37], Stochastic Subspace Identification (SSI) [38], Proper Orthogonal Decomposition (POD) [39–41] and spectral analysis [42, 43], and others.

Here, the limitations of the approaches based on single measured signal are summarized. First, they neglect the links to other nodes or generators and can not provide the spatial distributions of the mode. Second, they do not explore the relationships of the contained oscillation modes. Third, the calculated oscillatory parameters have no direct relations to the control strategy.

For the approaches based on the ensemble measurement matrix, it has been demonstrated that multi-Prony can extract the dominant oscillation mode and improve the identification accuracy. However, it is difficult to decide the correct model order of the system and the identification result is easily affected by noise [27]. SSI is proposed to analyze the LFO signal in [38]. It can capture the critical mode of a power system without requiring a considerable time . However, the single stochastic subspace can not deal with the nonlinear and nonstationary oscillation signals. Recently, several methods based on the multi-variable statistical data analysis have been proposed for spatiotemporal analysis of the measured matrix. Among of them, POD is a particularly useful tool to reduce the system model and to extract model information from measured data [39]. In [41], the data-driven framework which integrates the complex-empirical orthogonal function (EOF) analysis and the snapshots method is proposed to identify dynamic independent spatiotemporal patterns from time-synchronized data. Although the complex-EOF analysis can extract the principle oscillating components as well as their spatial and temporal parameters based on ensemble measurement matrix, it has some barriers in estimating the fluctuating part and explaining the negative frequency values. As for the mode shape estimation, the spectral correlation analysis is utilized to link the relationships between the mode shape and multi-signals [42]. In order to deal with the nonstationary signals, the spectral correlation analysis is updated by adding the weighted-updates algorithm [43]. The applications to simulated system and measured data from the Western North American power system have shown that the spectral correlation analysis can accurately track the mode shape and coherence over time in the presence of a major system configuration change.

Most of the methods based on the ensemble measurement matrix can extract the oscillation mode shape or similar results. However, they can not provide time-varying and nonlinear oscillatory parameters of each single measured signal. Furthermore, the calculated results can not be directly utilized to estimate the stability of the oscillation mode.

1.5 Challenges of Wide Area Dynamic Monitoring and Control

Although there are many theoretical researches on wide-area stability control, there are still many works that should be performed, in order to move forward the practical application of wide-area stability control in the large interconnected systems, which can be summarized as follows:

1. *Problem of interaction between local controllers and wide-area controllers.* In practice, large interconnected systems include many conventional stability control devices, as typically the PSS devices, using local control signals. These local controllers are generally designed to damp local oscillation modes. When the wide-area damping controller is applied in a power system, it may reduce the damping capability of local controllers on the local oscillations, although it can provide effective damping on LFO. Therefore, in order to guarantee the overall stability, it is significant to simultaneously optimize local and wide-area controllers. Till date, there have been many reports [44–46] on how to tune large numbers of local PSSs, but for the topic on how to tune plenty of local and wide-area controllers together, the related research is very few.

2. *Problem of control coordination among multiple wide-area damping controllers.* The centralized wide-area stability control, which utilizes various power system control devices (e.g., HVDC and FACTS controllers) and multi-channel wide-area control signals, is one of the trends of the smart transmission grids [12–48]. Since there are several LFO modes, this kind of centralized control can provide multiple damping on these dominant modes, which is good to improve the overall stability of the interconnected systems. However, it should be noted that the interaction among different control loops may occur if the coordination design is not carried out for multiple wide-area controllers. Such interactions may reduce the damping performance of multiple controllers, or it even may endanger the system stability. Therefore, the wide-area control coordination is also significant for the practical application of wide-area stability control.

3. *Problem of how to choose optimal control-input for multiple wide-area damping controllers.* For feedback control, the quality of the control performance depends directly on the proper selection of the feedback control-input.

Till date, some signal selection methods [49, 50] have been proposed, but these methods mainly select local signals for the power system controller. In fact, the WAMS application makes it feasible to select optimal control signal in the global range (including both local and wide-area range). Thus, comparing with the signal selection for a local controller, there are much more local and wide-area signal candidates for the selection of the control-input of the wide-area controller. But at present, the related research for the wide-area controller is very few. Besides this, once suitable signals cannot be selected for each of the multiple controllers, the interaction among these controllers is easily excited, which makes the design of coordinated multiple wide-area controllers more difficult.

4. *Delay effects of wide-area signals on the control performance of the wide-area controller.* For wide-area control, input signals needs to be received from the remote regions and furthermore wide-area control output signals have to be sent to other remote regions where the controllable devices are located. During signal processing, a time-delay is inevitably caused. Although the advanced synchronized phasor technology can control the time delay in around 300 ms [51, 52], many research results have shown that even a small time-delay can still lead to control failure of the wide-area controller. More seriously, under the effect of time delay of wide-area signals, the wide-area control may even destroy the stability of power systems. Therefore, an advanced control theory or time-delay compensation method should be considered to reduce or suppress the delay effect. At present, many controller design methods [53–57] have been proposed to handle with the time-delay, but most of them only concern a fixed time-delay. Under the assumption of specified time delays, these methods can reach good control performance. But in practice, the time-delay of the wide-area signals varies with time, and in such a case, these methods are not good to handle with time-varying delay, which limits the control performance of the wide-area controller.

5. *Problem of design and implementation for practical application of online wide-area monitoring and control.* As mentioned above, although there have been many researches on wide-area stability control, most of them only stay at the theoretical stage. To promote the practical application, there are still many problems to be considered. For examples, how to design an advanced control algorithm to suppress the delay effect of wide-area signal? How to establish a wide-area control network that can efficiently receive wide-area control-input signals, process the control algorithm and send wide-area control-output signals? How to embed the wide-area control system in power systems? These practical problems should be considered sufficiently for the practical application of wide-area stability control.

References

1. Vournas CD, Metsiou A, Nomikos BM (2009) Analysis of intra-area and interarea oscillations in South-Eastern UCTE interconnection. In: IEEE Power energy society general meeting, 2009
2. Lehner J, Kaufhold M, Treuer M, Weissbach T (2010) Monitoring of inter-area oscillations within the European interconnected network based on a wide area measuring system. In: IEEE PES transmission and distribution conference and exposition, 2010
3. Hashiguchi T, Mitani Y, Saeki O, Tsuji K, Hojo M, Ukai H (2004) Monitoring power system dynamics based on phasor measurements from demand side outlets developed in Japan Western 60 Hz system. In: IEEE power systems conference and exposition, 2004
4. Ota Y, Hashiguchi T, Ukai H, Sonoda M, Miwa Y, Takeuchi A (2007) Monitoring of interconnected power system parameters using PMU based WAMS. In: IEEE Lausanne Power Tech, 2007
5. Li P, Wu XC, Lu C, Shi JH, Hu J, He JB, Zhao Y, Xu AD (2009) Implementation of CSG's wide-area damping control system: overview and experience. In: IEEE/PES power system conference and exposition, 2009
6. Xie XR, Xin YZ, Xiao JY, Wu JT, Han YD (2006) WAMS applications in Chinese power systems. IEEE Power Energy Mag 4(1):54–63
7. Yuan ZY, Tao X, Zhang YC, Lang C, Markham PN, Gardner RM, Liu YL (2010) Inter-area oscillation analysis using wide area voltage angle measurements from FNET. In: IEEE power and energy society general meeting, 2010
8. Zhang YC, Markham P, Tao X, Lang C, Ye YZ, Wu ZY, Lei W, Bank J, Burgett J, Conners RW, Liu YL (2010) Wide-area frequency monitoring network (FNET) architecture and applications. IEEE Trans Smart Grid 1(2):159–167
9. IEEE Power Engineering Society (2003) IEEE Guide for Synchronous Generator Modeling Practices and Applications in Power System Stability Analyses. The Institute of Electrical and Electronics Engineers, Inc., 2003
10. Rogers G (2000) Power system oscillations. Kluwer, Norwell
11. Ke DP, Chung CY, Xue YS (2011) An eigenstructure-based performance index and its application to control design for damping inter-area oscillations in power systems. IEEE Trans Power Syst 26(4):2371–2380
12. Grebe E, Kabouris J, Lopez Barba S, Sattinger W, Winter W (2010) Low frequency oscillations in the interconnected system of continental Europe. In: Proceedings of IEEE· power energy society general meeting, 25–29 July 2010
13. Xie XR, Xin YZ, Xiao JY, Wu JT, Han YD (2006) WAMS applications in Chinese power systems. IEEE Power Energ Mag 4(1):54–63
14. Yuan ZY, Xia T, Zhang YC, Chen L, Markham PN, Gardner RM, Liu YL (2010) Inter-area oscillation analysis using wide area voltage angle measurements from FNET. In Proceedings of IEEE power energy society general meeting, 25–29 July 2010
15. Mao XM, Zhang Y, Guan L, Wu XC (2006) Coordinated control of inter-area oscillation in the China Southern power grid. IEEE Trans Power Syst 21(2):845–852
16. Mao XM, Zhang Y, Guan L, Wu XC, Zhang M (2008) Improving power system dynamic performance using wide-area high-voltage direct current damping control. IET Gener Transm Distrib 2(2):245–251
17. Rogers G (2000) Power system oscillations. Kluwer Academic Publisher, Norwell. ISBN 0-7923-7712-5
18. Liu F (2011) Robust design of FACTS wide-area damping controller considering signal delay for stability enhancement of power system. A dissertation for PhD degree, Waseda University
19. Martins N, Lima LTG (1990) Determination of suitable locations for power system stabilizers and static var compensators for damping electromechanical oscillations in large scale power systems. IEEE Trans Power Syst 5(4):1455–1469

20. Wang L, Semlyen A (1990) Application of sparse eigenvalue techniques to the small signal stability analysis of large power systems. IEEE Trans Power Syst 5(2):635–642
21. Deng JX, Tu J, Chen WH (2007) Identification of critical low frequency oscillation mode in large disturbances. Power Syst Technol 31(7):36–41
22. Prabha K (2004) Power system stability and control. McGraw-Hill, New York
23. Huang Y, Xu Z, Pan WL (2005) Analysis method for low frequency oscillation in east china power grid based on power system simulation software PSS/E. Power Syst Technol 29 (23):11–17
24. Huang Z, Zhou N, Tuffner FK, et al (2010) MANGO-modal analysis for grid operation: a method for damping improvement through operating point adjustment. Pacific Northwest National Laboratory operated by Battelle for the United States Department of energy under contract DE-AC05-76RL01830
25. Han S (2011) Research on dynamic characteristic analysis methods for inter-area oscillations based on WAMS. A dissertation for PhD degree, Zhejiang University
26. Poon KP, Lee KC (1988) Analysis of transient stability swings in large interconnected power systems by Fourier transformation. IEEE Trans Power Syst 3(4):1573–1581
27. Hauer J, Demeure C, Scharf L et al (1990) Initial results in prony analysis of power system response signals. IEEE Trans Power Syst 5(1):80–89
28. Xiao JY, Xie XR, Hu ZX (2004) Improved prony method for online identification of low-frequency oscillations in power systems. J Tsinghua Univ (Sci Technol) 44(7):883–887
29. Yuan Y, Sun YZ, Cheng L et al (2010) Power system low frequency oscillation monitoring and analysis based on multi-signal online identification. Sci Chin Ser E-Tech Sci 53(9):2589–2596
30. Wiltshire RA, Ledwich G, OShea P (2007) A Kalman filtering approach to rapidly detecting modal changes in power systems. IEEE Trans Power Syst 22(4):1698–1706
31. Messina AR, Vittal V, Ruiz-Vega D et al (2006) Interpretation and visualization of wide-area PMU measurements using Hilbert analysis. IEEE Trans Power Syst 21(4):1761–1763
32. Messina AR (2009) Inter-area oscillations in power systems: a nonlinear and non-stationary perspective. Springer Press, Berlin. ISBN 978-0-387-89529-1
33. Li TY, Gao L, Zhao Y (2006) Analysis of low frequency oscillations using HHT Method. Proc CSEE 26(14):24–29
34. Rueda JL, Juarez CA, Erlich I (2011) Wavelet-based analysis of power system low-frequency electromechanical oscillations. IEEE Trans Power Syst 99(2):1–11
35. Jukka T (2011) A wavelet-based method for estimating damping in power systems. A dissertation for PhD degree, Aalto University
36. Ma YF, Zhao SQ, Gu XP (2007) Reduced order identification of low-frequency oscillation transfer function and PSS design based on improved multi-signal prony algorithm. Power Syst Technol 31(17):16–20
37. Wang H, Su XL (2011) Generation unit correlativity-based prony analysis on multi-signal classification of low-frequency oscillation. Power Syst Technol 35(6):128–133
38. Cai GW, Yang DY, Jiao Y et al (2009) Power system low frequency oscillations analysis and parameter determination of adaptive PSS based on stochastic subspace identification. In: Proceedings in power and energy engineering conference (APPEEC), Chengdou, China, 27–31 Mar 2009
39. Messina AR, Vittal V (2007) Extraction of dynamic patterns from wide-area measurements using empirical orthogonal functions. IEEE Trans Power Syst 22(2):682–692
40. Messina AR, Esquivel P, Lezama F (2011) Wide-area PMU data monitoring using spatio-temporal statistical models. In: Power system conference and exposition (PSCE), IEEE/PES, 20–23 Mar 2011
41. Esquivel P (2010) Extraction of dynamic patterns from wide area measurements using empirical orthogonal functions. A dissertation for PhD degree, CINVESTAV del IPN Unidad Guadalajara, Aug 2010
42. Trudnowski DJ (2008) Estimating electromechanical mode shape from synchro-phasor measurements. IEEE Trans Power Syst 23(5):1188–1195

43. Tuffner FK, Dosiek L, Pierre JW et al (2010) Weighted update method for spectral mode shape estimation from PMU measurements. In: IEEE power engineering society general meeting, 25–29 July 2010
44. Do Bomfim ALB, Taranto GN, Falcao DM (2000) Simultaneous tuning of power system damping controllers using genetic algorithms. IEEE Trans Power Syst 15(1):163–169
45. Abdel-Magid YL, Abido MA, Al-Baiyat S, Mantawy AH (1999) Simultaneous stabilization of multimachine power systems via genetic algorithms. IEEE Trans Power Syst 14(4):1428–1439
46. Cai LJ, Erlich I (2005) Simultaneous coordinated tuning of PSS and FACTS damping controllers in large power systems. IEEE Trans Power Syst 20(1):294–300
47. Jiang ZH, Li FX, Qiao W, Sun HB, Wan H, Wang JH, Xia Y, Xu Z, Zhang P (2009) A vision of smart transmission grids. In: IEEE proceedings of power and energy society general meeting, Jul 2009, pp 1–10
48. De La Ree J, Centeno V, Thorp JS, Phadke AG (2010) Synchronized phasor measurement applications in power systems. IEEE Trans Smart Grid 1(1):20–27
49. Farsangi MM, Song YH, Lee KY (2004) Choice of FACTS device control inputs for damping inter-area oscillations. IEEE Trans Power Syst 19(2):1135–1143
50. Farsangi MM, Nezamabadi-pour H, Song YH, Lee KY (2007) Placement of SVCs and selection of stabilizing signals in power systems. IEEE Trans Power Syst 22(3):1061–1071
51. Wu HX, Tsakalis KS, Heydt GT (2004) Evaluation of time delay effects to wide-area power system stabilizer design. IEEE Trans Power Syst 19(4):1935–1941
52. Stahlhut JW, Browne TJ, Heydt GT, Vittal V (2008) Latency viewed as a stochastic process and its impact on wide area power system control signals. IEEE Trans Power Syst 23(1):84–91
53. Dotta D, e Silva AS, Decker IC (2009) Wide-area measurement-based two-level control design considering signals transmission delay. IEEE Trans Power Syst 24(1):208–216
54. Majumder R, Pal BC, Dufour C, Korba P (2006) Design and real-time implementation of robust FACTS controller for damping inter-area oscillation. IEEE Trans Power Syst 21 (2):809–816
55. Majumder R, Chaudhuri B, Pal BC (2005) A probabilistic approach to model-based adaptive control for damping of interarea oscillations. IEEE Trans Power Syst 20(1):367–374
56. Chaudhuri B, Pal BC (2004) Robust damping of multiple swing modes employing global stabilizing signals with a TCSC. IEEE Trans Power Syst 19(1):499–506
57. Chaudhuri NR, Ray S, Majumder R, Chaudhuri B (2010) A new approach to continuous latency compensation with adaptive phasor power oscillation damping controller (POD). IEEE Trans Power Syst 25(2):939–946

Chapter 2
Theoretical Foundation of Low-Frequency Oscillations

In this chapter, the theoretical foundation of low-frequency oscillation (LFO) is introduced. With the increasing utilizations of measurement devices throughout the system, especially the construction of WAMS, the methods which are based on the measured data on-time or off-time have been paid more attention in last few years. First, the basic principles and research techniques of LFO are introduced. Then two kinds of methods based on system model and measured information are presented in detail, such as the eigenvalue method, the discrete Fourier transform (DFT), Hilbert–Huang transform (HHT), and so on.

2.1 The Basic Principles of Low-Frequency Oscillation

From the practical operation of power system, the reasons which may trigger the LFO phenomenon are summarized as follows: the chain structure of transmission system; the weak connection and power imbalance among different areas; shortage reserve capacity; and interactions of different control devices.

Based on the existing publications, the negative damping theory, resonance mechanism, and nonlinear theory have been proposed to explain the principles of LFOs: (1) negative damping theory is based on the linear theory and it has been widely accepted by the community. The regulations of excitation system on the generator can produce the additional negative damping. It may offset the positive damping of generator. Then, the total damping ratio is nearly to zero or even negative. Under this condition, the oscillation would be gradually amplified and cause the out of step of the generators if the system is disturbed by a small disturbance. Improving the damping by compensating the phase is the basic design idea of the LFO controller [1, 2]; (2) in some actual power systems, not all the LFO phenomena can be interpreted by the negative damping theory. Many researchers have pointed out that constant amplitude power oscillation will occur when the disturbance frequency of prime mover power is nearly to the natural frequency of power system, even if the damping is strong enough [3]. In [4], it proposed an identification method to locate disturbance source by the energy conversion principles based on the comparison of negative damping theory and forced mechanism; (3) the nonlinear mechanism means that the stable structure of power system can produce nonlinear and irregular oscillations when the parameters and disturbance are in the special range [5, 6]. It has big difference from the linear theory. However, this principle still stays in the period of theoretical study because of the limitation of mathematical tool. It should be noted that the principle of LFO in this paper is based on the negative damping theory.

Up to now, there are many different classification methods according to different criteria. Usually, LFO belongs to the branch of rotor angle stability. According to widely accepted results, the classification of LFO is divided into two forms: one is the local oscillation mode; the other is the inter-area oscillation mode [1, 7].

2.1.1 Local Mode

Local oscillation mode involves in the oscillation between the single generator or a group of generators and the rest of generators in the local area. In general, the oscillation frequency of this mode is in the range of 1–2.5 Hz. Due to this definition, it can be concluded that the influence of the local mode is just in the specific area and it can be controlled expediently by the PSS [1].

To investigate the physical nature of low-frequency electromechanical power oscillations in the time domain, a model of a typical two-area power system was created. This two-area system was created by Ontario Hydro for a research report commissioned by the Canadian Electrical Association [8, 9]. This system was designed to exhibit the different types of oscillations that occur in an interconnected system [8, 9]. This two-area system can be considered as a useful tool for the study of the electromechanical oscillations in the GB power system. For this purpose Area 1 represents the Scottish power system and Area 2 represents the English power system. A single line diagram of the two-area system is shown in Fig. 2.1.

This simple model shows the electromechanical oscillations that are inherent in the two-area system. There are three possible electromechanical modes of oscillations in this system. There are two local modes, one in which generator 1 swings

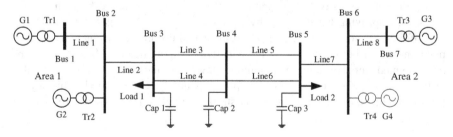

Fig. 2.1 A typical two-area system

against generator 2, and another in which generator 3 swings against generator 4. In addition, there is also one inter-area mode, in which the generators in Area 1 swing against the generators in Area 2. In this section, nonlinear simulations will be used to give an insight into the nature of these different types of electromechanical oscillations. In these nonlinear simulations, the different modes of oscillation are initiated using a range of different disturbances.

To investigate the nature of the local mode in Area 1, changes in the mechanical torque of the generators in that area were simulated. To properly investigate the behavior of the local mode in Area 1 it is important to minimize the excitation of the inter-area mode during these simulations. To achieve this goal, equal and opposite step changes in the mechanical torque of the two generators in Area 1 were simulated simultaneously. For example, a change of −0.01 p.u. in the mechanical torque of generator 1 is simulated, and then a corresponding change of 0.01 p.u. is made in the mechanical torque of generator 2. The response of the generators, in terms of speed, to this pair of small disturbances in Area 1 is presented in Fig. 2.2.

In Area 1, the rotor speed changes of generator 1 and 2 were in anti-phase, i.e, generator 1 oscillated against generator 2 in the local mode. This local mode dominated the oscillation for approximately 7 s, at which time the generators began to swing together in the inter-area mode. The generators in Area 2 experienced oscillations with lower amplitude than those seen in Area 1. These oscillations were in phase with one another and are driven by the inter-area mode, the local mode in Area 2 was not observed here. These simulation results show that the frequency of the local oscillation mode in Area 1 is approximately 1 Hz.

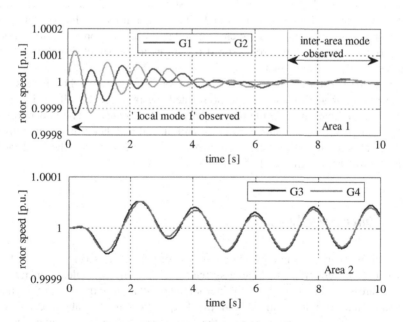

Fig. 2.2 Generator rotor speed responses to the disturbances occurred in Area 1

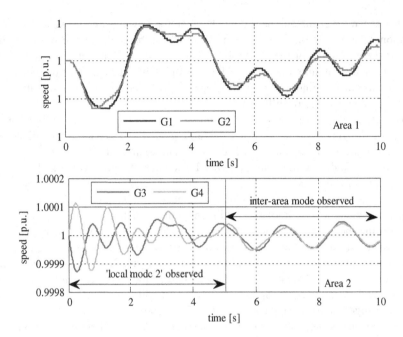

Fig. 2.3 Generator rotor speed responses to the disturbances occurred in Area 2

The same method used to excite the local mode in Area 1 is used to excite the local mode in Area 2. An equal and opposite step change of the mechanical torque of the generators in Area 2 was simulated. The change in the mechanical torque of generator 3 was −0.01 p.u. and the change in the mechanical torque of generator 4 was 0.01 p.u. The generator rotor speed responses to the small disturbances that occurred in Area 2 are shown in Fig. 2.3.

For a small disturbance in Area 2, generator 3 immediately began to swing against generator 4; this local mode dominated the response for about 5 s, after which time the inter-area oscillatory mode began to dominate. The generators in Area 1 were driven by the inter-area mode and moved together with oscillations of much lower amplitude than those seen in Area 2. The frequency of the local mode in Area 2 was approximately 1 Hz.

2.1.2 Inter-Area Mode

Inter-area oscillation mode represents one part of generators swinging against the other part of generators in wide-area systems. Because the equivalent generator in different areas has bigger constant of inertia, the oscillation frequency of this mode is smaller than the local mode and the frequency range is about 0.1–1 Hz. Compared with the local mode, the inter-area mode has some special

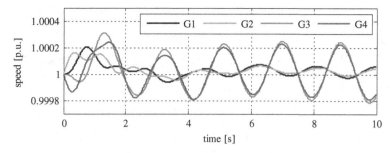

Fig. 2.4 Generator rotor speed oscillations dominated by inter-area mode

characteristics: the lasting time is longer, the impact is more slowly and widely, the damping value is smaller, and it is more difficult to be controlled [1].

The inter-area mode can be directly provoked by changing the mechanical torque of one generator in each of the different areas. In this case, the mechanical torque of generator 2 was increased by 0.01 p.u., while the mechanical torque of generator 4 was reduced by 0.01 p.u. The generator speed responses to these small disturbances are shown in Fig. 2.4.

As Fig. 2.4 presents, the inter-area mode dominated the response of the generator rotor speeds to these disturbances. The generators in Area 1 began to swing against the generators in Area 2 immediately after the disturbances, and the magnitudes of the speed change of the generators in Area 2 were larger than the magnitudes of the speed change of the generators in Area 1. Initially, the oscillations in Area 1 were strongly influenced by the local mode. This is evident as for the first 4 s the generators in Area 1 oscillated against one another while also moving together in the inter-area mode. The frequency of the inter-area mode was approximately 0.5 Hz. The inter-area mode was not damped by any external control and the amplitude of the inter-area oscillation was seen to increase.

For obtaining more information about inter-area oscillations, the responses of the system frequency and the inter-area active power flow to the disturbances were analyzed. Figure 2.5 presents the system frequency response to the disturbances measured in Area 1 (bus 3) and Area 2 (bus 5). The oscillations in the frequency deviation in Area 1 were approximately in anti-phase to the oscillations in the frequency deviation in Area 2, which is consistent with the changes seen in the generator rotor speeds in Fig. 2.4.

Figure 2.6 shows the active power flow over line 3 after the disturbances. The oscillatory power flow on line 3 is purely driven by the inter-area mode, with no influence from the local modes. This occurs because the physical mechanism behind electromechanical oscillations is the active power exchange between the generators that are involved in the oscillatory mode. Therefore, as line 3, like all of the inter-tie lines, only carries power between the two areas, then only the inter-area oscillations, and not the local mode oscillations, will be seen on these lines.

The power exchange driving the local modes in Areas 1 and 2 occurs along lines 1 and 8, respectively. Therefore, the power flow associated with these local

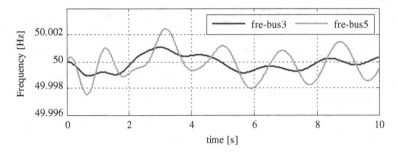

Fig. 2.5 System frequency responses in inter-area mode

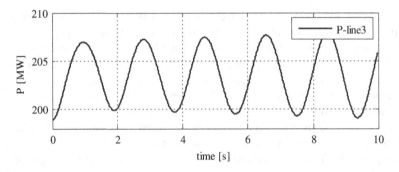

Fig. 2.6 Oscillatory active power flow on transmission line 3

oscillation modes will only be seen on these lines. To demonstrate this, the power flow on line 1 during the disturbances is shown in Fig. 2.7. The variation of the power flow on line 1 shows the difference between the power flow that supports the local mode and the power flow that supports the inter-area mode. This can be seen by comparing the power flow during the first four seconds after the disturbances, where the local mode dominates, with the power flow during the next six seconds, where the inter-area mode dominates.

To further examine the characteristics of the inter-area oscillations, a three-phase short-circuit fault was simulated on bus 4, the mid-point of the inter-area lines that connect the two areas. A short circuit represents a much larger disturbance to the system than the small mechanical power changes simulated in the previous sections and as such will offer greater insight into the behavior of the oscillatory modes. The transient fault occurred at 0.1 s and was cleared after at 0.2 s. The response of the rotor speed of each generator to the disturbance is presented in Fig. 2.8, and the active power transfer over one of the inter-tie line (line 3) is shown in Fig. 2.9.

As seen from Figs. 2.8 and 2.9, after the system recovered from the transient fault, the generators in Area 1 started to oscillate against the generators in Area 2 in inter-area mode around the new system equilibrium point. The inter-area mode was clearly visible in the generator rotor speed responses and the oscillatory active power flow on the inter-tie line.

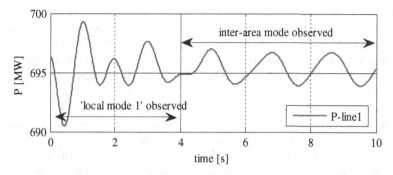

Fig. 2.7 Oscillatory active power flow on transmission line 1

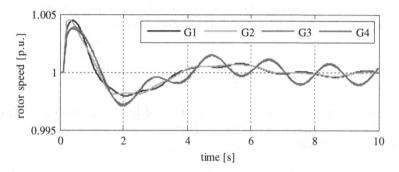

Fig. 2.8 Responses of the generator rotor speeds to the large disturbance

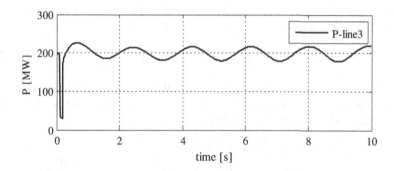

Fig. 2.9 Active power transfer over the tie line after the disturbance

2.2 Techniques Based on System Model

In Sect. 2.1, a number of nonlinear simulations were performed to show the physical nature behind power system oscillations. In the nonlinear simulations the transients were induced using small disturbances (e.g., a 1 % change of mechanical torque), and the system responses were essentially linear. Although the initial system response to the three-phase transient fault was nonlinear, the response quickly settled into post-fault oscillations that are essentially linear around the post-fault equilibrium point. This means that, for a study of electromechanical oscillations, the system model can be linearized around the steady-state point. The linearization of the system model provides an excellent opportunity for modal analysis [8]. Modal analysis can be used to perform a wide range of tasks, such as determining the oscillatory modes, the sources of the oscillatory modes, and the parameters needed for designing oscillation controllers.

In this section, the modal analysis methods will be introduced. This analysis tool will then be used to explain the physical phenomenon of inter-area oscillations presented in the Sect. 2.1. Generally speaking, this method based on the system model or device parameters can be classified into two kinds: one is the linear methods, such as the eigenvalue analysis; the other is nonlinear methods, for example, time-domain simulation and normal form analysis.

2.2.1   Linearization of the State Equation

If there are a set of state variables x_0 and a set of inputs u_0 with which all the derivatives $\dot{x}_1, \dot{x}_2, \cdots, \dot{x}_n$ are simultaneously zero, as represented by the (2.1), we can say that the system is at a steady state [10].

$$\dot{x}_0 = f(x_0, u_0) \tag{2.1}$$

If a stable system is disturbed from steady state by a small disturbance, i.e., Δx and Δu, it will eventually come to rest at a new steady state. This transition will still satisfy (2.1), and hence we have

$$\dot{x} = f[(x_0 + \Delta x), (u_0 + \Delta u)] \tag{2.2}$$

As the disturbance is assumed to be very small, the nonlinear equation $f(x, u)$ can be approximated using a Taylor series expansion in terms of Δx and Δu. If we only consider the first-order terms of this expansion we will have

$$\dot{x}_i = \dot{x}_{i0} + \Delta \dot{x}_i = f_i[(x_0 + \Delta x), (u_0 + \Delta u)]$$
$$= f_i(x_0, u_0) + \frac{\partial f_i}{\partial x_1} \Delta x_1 + \cdots + \frac{\partial f_i}{\partial x_n} \Delta x_n + \frac{\partial f_i}{\partial u_1} \Delta u_1 + \cdots + \frac{\partial f_i}{\partial u_r} \Delta u_r \tag{2.3}$$

Since $\dot{x}_{i0} = f_i(x_0, u_0)$,we have

$$\Delta \dot{x}_i = \frac{\partial f_i}{\partial x_1} \Delta x_1 + \cdots + \frac{\partial f_i}{\partial x_n} \Delta x_n + \frac{\partial f_i}{\partial u_1} \Delta u_1 + \cdots + \frac{\partial f_i}{\partial u_r} \Delta u_r, \quad i = 1, 2, \ldots, n \tag{2.4}$$

in the same way based on (2.2), we have

$$\Delta y_i = \frac{\partial g_i}{\partial x_1}\Delta x_1 + \cdots + \frac{\partial g_i}{\partial x_n}\Delta x_n + \frac{\partial g_i}{\partial u_1}\Delta u_1 + \cdots + \frac{\partial g_i}{\partial u_r}\Delta u_r, \quad j = 1, 2, \ldots, m$$

$$(2.5)$$

Therefore, the linearized system state equations around the equilibrium point are given as [10]

$$\Delta \dot{x} = A\Delta x + B\Delta u \tag{2.6}$$

$$\Delta y = C\Delta x + D\Delta u \tag{2.7}$$

with

$$A = \begin{bmatrix} \frac{\partial f_1}{\partial x_1} & \cdots & \frac{\partial f_1}{\partial x_n} \\ \vdots & \vdots & \vdots \\ \frac{\partial f_n}{\partial x_1} & \cdots & \frac{\partial f_n}{\partial x_n} \end{bmatrix}, \quad B = \begin{bmatrix} \frac{\partial f_1}{\partial u_1} & \cdots & \frac{\partial f_1}{\partial u_r} \\ \vdots & \vdots & \vdots \\ \frac{\partial f_n}{\partial u_1} & \cdots & \frac{\partial f_n}{\partial u_r} \end{bmatrix},$$

$$C = \begin{bmatrix} \frac{\partial g_1}{\partial x_1} & \cdots & \frac{\partial g_1}{\partial x_n} \\ \vdots & \vdots & \vdots \\ \frac{\partial g_m}{\partial x_1} & \cdots & \frac{\partial g_m}{\partial x_n} \end{bmatrix}, \quad D = \begin{bmatrix} \frac{\partial g_1}{\partial u_1} & \cdots & \frac{\partial g_1}{\partial u_r} \\ \vdots & \vdots & \vdots \\ \frac{\partial g_m}{\partial u_1} & \cdots & \frac{\partial g_m}{\partial u_r} \end{bmatrix}.$$

where Δx is the state vector of length n; Δu is input disturbance vector of length r; Δy is the output vector of length m; A is the state matrix of size $n \times n$; B is the input matrix of size $n \times r$; C is the output matrix of size $m \times n$; and D is a matrix of size $m \times r$, which defines the proportion of the input which directly influences the output, Δy.

2.2.2 Calculation of Eigenvalues and Eigenvectors

The following equation is defined as the characteristic equation of matrix A [10]:

$$\det(\lambda I - A) = 0 \tag{2.8}$$

The n solutions of the characteristic equation $\lambda = [\lambda_1, \lambda_2, \ldots, \lambda_n]$ are the eigenvalues of A. The eigenvalues may be real or complex. If matrix A is real, then the complex eigenvalues occur in conjugate pairs.

For any eigenvalue λ_i, a n-column vector φ_i satisfied

$$A\varphi_i = \lambda_i \varphi_i \quad i = 1, 2, \ldots, n \tag{2.9}$$

is the right eigenvector of the A matrix associated with the eigenvalue λ_i. The right eigenvector ϕ_i has the form

$$\varphi_i = \begin{bmatrix} \varphi_{1i} \\ \varphi_{2i} \\ \cdots \\ \varphi_{ni} \end{bmatrix} \tag{2.10}$$

The right eigenvector describes how each mode of oscillation is distributed among the system states. In other words, it indicates on which system variables the mode is more observable [11]. The magnitudes of the elements of φ_i give the extent of the behaviors of the n state variables in the ith mode, and the angles of elements give the phase displacements of the state variables with regard to the mode. Thus, the right eigenvector is called mode shape.

Similarly, if a n-row vector φ_i satisfies

$$\psi_i A = \lambda_i \psi_i \quad i = 1, 2, \ldots, n \tag{2.11}$$

it is called the left eigenvector of the A matrix associated with the eigenvalue λ_i. The left eigenvector has the form

$$\psi_i = [\psi_{i1} \ \psi_{i2} \ldots \psi_{in}] \tag{2.12}$$

The left eigenvector ψ_i can be used to identify which combination of state variables displays only the ith mode. Thus, the kth element of the right eigenvector φ_i measures the activity of the variable x_k in the ith mode, and the kth element of the left eigenvector ψ_i weights the contribution of this activity to the ith mode.

2.2.3 Determination of Oscillation Parameters

The damping ratio ξ determines the decaying rate of the amplitude of an oscillation. The eigenvalues λ of A matrices can be obtained by solving the root of the characteristic equation. For a complex–conjugate pair of eigenvalues

$$\lambda = \sigma \pm j\omega \tag{2.13}$$

the corresponding damping ratio and oscillation frequency f can be defined as follows:

$$\xi = \frac{-\sigma}{\sqrt{\sigma^2 + \omega^2}} \tag{2.14}$$

$$f = -\frac{\omega}{2\pi} \tag{2.15}$$

If λ_i is an eigenvalue of A, v_i and w_i are nonzero column and row vectors, respectively, such that the following relations hold:

$$Av_i = \lambda_i v_i \quad i = 1, 2, \ldots, n \tag{2.16}$$

$$w_i A = \lambda_i w_i \quad i = 1, 2, \ldots, n \tag{2.17}$$

where the vectors v_i and w_i are known as the right and left eigenvectors of matrix A. And they are henceforth considered normalized such that

$$w_i \cdot v_i = 1 \tag{2.18}$$

Then the participation factor P_{ki} (the kth state variable x_k in the ith eigenvalue λ_i) can be given as

$$P_{ki} = |v_{ki}||w_{ki}| \tag{2.19}$$

where w_{ki} and v_{ki} are the ith elements of w_k and v_k, respectively.

2.2.4 Brief Summary of System Model Analysis Techniques

Model-based method has been widely utilized to extract the modal information as well as to identify the transfer function and to design the controller. Generally speaking, this method based on the system model or device parameters can be classified into two kinds: one is the linear methods, such as the eigenvalue analysis; the other is nonlinear methods, for example, time-domain simulation and normal form analysis. The advantages of these approaches are that they allow the investigation of the dynamic behavior of the system under different modes and all the characteristic patterns leading to complicated phenomena can be detected. Moreover, the nonlinear behaviors of other controllers are also taken into account simultaneously. However, there are also many limitations for the application of the time-domain simulation: (1) it is impossible to trigger all the oscillation modes because the disturbance location and style are chosen specifically; (2) it is difficult to determine the oscillation characteristics only based on the limited time-domain responses; (3) time-domain simulation model is based on the detailed parameters of all elements, and it is impossible to get these parameters accurately because of the benefits of manufacturers; and (4) longer simulation time and heavy computation load do not make it fulfill the requirements of online application. According to the features of eigenvalue analysis and time-domain simulation, both methods should be utilized mutually and in a complementary manner.

2.3 Techniques Based on Measured Information

A large collection of methods has been investigated for extracting model information from measured system responses. In this section, these methods are divided into two groups according to the number of disposed signals. The first kind of approach is presented to calculate the oscillatory parameters of single measured signal, such as discrete Fourier transform (DFT), Hilbert–Huang transform (HHT), wavelet transform (WT), and so on. The second kind of method is presented to extract the dynamic oscillation characteristic based on the ensemble measurement matrix, such as Prony method, multi-Prony technique, and so on.

2.3.1 Discrete Fourier Transform

Discrete Fourier transform (DFT) is a form of Fourier analysis that is applicable to the uniformly spaced samples taken from an input signal $x(t)$ of a continuous function. In particular, the DFT is the primary tool of digital signal processing and related fields. A key enabling factor for its widely applications is the fact that the DFT can be computed efficiently in practice using a fast Fourier transform (FFT) algorithm. The FFT is calculated at discrete steps in the frequency domain, just as the input signal is sampled at discrete instants in the time domain [12].

Consider the process of selecting N samples: $x(k\Delta T)$ with $\{k = 0, 1, \ldots, N - 1\}$, ΔT being the sampling interval. This is equivalent to multiplying the sampled data train by a "windowing function" $\omega(t)$, which is a rectangular function of time with unit magnitude and a span of $N\Delta T$. With the choice of samples ranging from 0 to $N - 1$, the windowing function can be viewed as starting at $-\Delta T/2$ and ending at $(N - 1/2)\Delta T$.

The collection of signal samples fall in the data window: $x(k\Delta T)$ with $\{k = 0, 1, \ldots, N - 1\}$. These samples can be viewed as being obtained by the multiplication as follows:

$$y(t) = x(t)\omega(t)\delta(t) = \sum_{k=0}^{N-1} x(k\Delta T)\delta(t - k\Delta T) \tag{2.20}$$

The convolution of FFTs of the three functions finally get the FFT of the sampled windowed function $y(t)$.

In order to obtain the DFT of $y(t)$, the FFT of $y(t)$ is to be sampled in the frequency domain. The discrete steps in the frequency domain are multiples of $1/T_0$, where T_0 is the span of the windowing function. The frequency sampling function $\Phi(f)$ is given by

$$\Phi(f) = \sum_{n=-\infty}^{\infty} \delta\left(f - \frac{n}{T_0}\right) \tag{2.21}$$

and its inverse FFT is

$$\Phi(t) = T_0 \sum_{n=-\infty}^{\infty} \delta(t - nT_0) \tag{2.22}$$

It is essential to multiply the FFT $Y(f)$ with $F(f)$ to obtain the samples in the frequency domain. To obtain the corresponding time-domain function $x'(t)$, a convolution in the time domain of $y(t)$ and $\Phi(t)$ will be required:

$$
\begin{aligned}
x'(t) = y(t)\Phi(t) &= \left[\sum_{k=0}^{N-1} x(k\Delta T)\delta(t - k\Delta T)\right]\left[T_0 \sum_{n=-\infty}^{\infty} \delta(t - nT_0)\right] \\
&= T_0 \sum_{n=-\infty}^{\infty} \left[\sum_{k=0}^{N-1} x(k\Delta T)\delta(t - k\Delta T - nT_0)\right]
\end{aligned}
\tag{2.23}
$$

Through $N - 1$ and the sampling in frequency domain, the original N samples in time domain are transformed to an infinite train of N samples with a period T_0. Note that we may still consider this function to be an approximation of $x(t)$ although the original function $x(t)$ was not periodic and the function is $x'(t)$.

The FFT of the periodic function $x'(t)$ is a sequence of impulse functions in frequency domain of the FFT. Thus

$$x'(f) = \sum_{n=-\infty}^{\infty} \alpha_n \delta\left(f - \frac{n}{T_0}\right) \tag{2.24}$$

with

$$\alpha_n = \frac{1}{T_n} \int_{-T_0/2}^{T_0/2} x'(t) e^{-\frac{j2\pi n}{T_0}} dt, \quad n = 0, \pm 1, \pm 2 \dots \tag{2.25}$$

Substituting for $x'(t)$ in the above expression for α_n,

$$\alpha_n = \frac{1}{T_0} \int_{-T_0/2}^{T_0/2} T_0 \sum_{m=-\infty}^{\infty} \left[\sum_{k=0}^{N-1} x(k\Delta T)\delta(t - k\Delta T - mT_0) e^{-\frac{j2\pi n}{T_0}} dt\right], \quad n = 0, \pm 1, \pm 2 \dots$$

$$\tag{2.26}$$

The index m designates the periodic sequences of the FFT of the windowed function. Since the limits on the integration span one period only, we may remove the summation on m, and set $m = 0$, thus using only the samples over the period. (2.22) then becomes

$$\alpha_n = \int_{-T_0/2}^{T_0/2} \sum_{k=0}^{N-1} x(k\Delta T)\delta(t - k\Delta T)e^{-\frac{j2\pi n}{T_0}}dt, \quad n = 0, \pm 1, \pm 2 \dots \quad (2.27)$$

or

$$\alpha_n = \sum_{k=0}^{N-1} \int_{-T_0/2}^{T_0/2} x(k\Delta T)\delta(t - k\Delta T)e^{-\frac{j2\pi n}{T_0}}dt = \sum_{k=0}^{N-1} x(k\Delta T)\delta(t - k\Delta T)e^{-\frac{j2\pi nk\Delta T}{T_0}} \quad (2.28)$$

Since there are N samples in the data window T_0, $N\Delta T = T_0$. Therefore,

$$\alpha_n = \sum_{k=0}^{N-1} x(k\Delta T)e^{-\frac{j2\pi nk}{N}} \quad (2.29)$$

with

$$n = 0, \pm 1, \pm 2 \dots$$

Note that there are only N distinct coefficients α_n although the index n goes over all positive and negative integers. Thus, α_{N+1} is the same as α_1 and the FFT $x'(f)$ has only N distinct values corresponding to frequencies $f = n/T_0$, with n ranging from 0 through $N - 1$.

2.3.2 Prony Algorithm and Multi-Prony

Prony method is a linear combination of complex exponential functions to describe the mathematical model of equal interval sampling data, often referred to as the Prony model, and gives the approximate solution algorithm. Prony analysis has been shown to be a feasible technique to model a linear combination of damped complex exponentials to signals that are uniformly sampled.

There are three basic steps in the Prony algorithm:

Step 1 Determine the linear prediction parameters that fit the available data.
Step 2 Solve the roots of the polynomial formed in *step 1*, and find the prediction coefficient that will yield the estimates of damping and sinusoidal frequencies of each of the exponential terms.

Step 3   With the solution obtained in *step 2*, a second linear equation will
        obtained.

Then, the estimate of the exponential amplitude and sinusoidal initial phase can
be received. Prony algorithm and multi-Prony estimate parameters of modes
associated with a signal, or certain set of signals [13], such as mode frequency,
amplitude, phase, and damping ratio. The Prony method calculates or estimates a
linear exponential function of the form (2.30) by approximating in the least square
sense to a certain set of equally sampled discrete data.

$$\hat{y}(t) = \sum_{i=1}^{m} A_i e^{\sigma_i t} \cos(2\pi f_i t + \varphi_i) \quad \text{for } t \geq 0 \tag{2.30}$$

Equation (2.30) may be written in a complex exponential form as

$$\hat{y}(t) = \sum_{i=1}^{m} B_i e^{\lambda_i t} + B_i^* e^{\lambda_i^* t} \quad \text{for } t \geq 0 \tag{2.31}$$

where m is the number of modes, $B_i = \frac{A_i}{2} e^{j\varphi_i}$, $\lambda_i = \sigma_i + j\omega_i$ and * indicates complex
conjugate. Moreover, (2.31) could be expressed in a simplified manner as

$$\hat{y}(t) = \sum_{i=1}^{p} B_i e^{\lambda_i t} \quad \text{for } t \geq 0 \tag{2.32}$$

where p is the number of the estimated eigenvalues.

Multi-Prony analysis is just a vector–matrix extension of Prony analysis, considering multiple outputs of the form $y(t) = cx(t)$ at the same time [14].

Consider also that, for (2.31) or (2.32), the B_i and the λ_i are distinct. Let $y(t)$ be
N samples evenly spaced by Δt such that

$$y(t_k) = y(k) \quad \text{for } k = 0, 1, \ldots, N - 1 \tag{2.33}$$

The Prony estimated output signal at time $t_k = k$ will be

$$\hat{y}(t) = \sum_{i=1}^{p} B_i e^{\lambda_i k \Delta t} \quad \text{for } k = 0, 1, \ldots, N - 1 \tag{2.34}$$

For convenience, define $z_i = e^{\lambda_i \Delta t}$, then

$$\hat{y}(t) = \sum_{i=1}^{p} B_i z_i^k \quad \text{for } k = 0, 1, \ldots, N - 1 \tag{2.35}$$

As can be seen, the objective with Prony method is to find the values of B_i and z_i that produce

$$\hat{y}(k) = y(k) \quad \text{for all } k \tag{2.36}$$

Assuming that (2.36) is true for all k, set

$$\begin{aligned}
y(0) &= B_1 + B_2 + \cdots + B_p \\
y(1) &= B_1 z_1 + B_2 z_2 + \cdots + B_p z_p \\
y(2) &= B_1 z_1^2 + B_2 z_2^2 + \cdots + B_p z_p^2 \\
&\;\;\vdots \qquad\quad \vdots \\
y(N-1) &= B_1 z_1^{N-1} + B_2 z_2^{N-1} + \cdots + B_p z_p^{N-1}
\end{aligned} \tag{2.37}$$

In matrix form,

$$\begin{bmatrix} y(0) \\ y(1) \\ y(2) \\ \vdots \\ y(N-1) \end{bmatrix} = \begin{bmatrix} z_1^0 & z_2^0 & \cdots & z_p^0 \\ z_1^1 & z_2^1 & \cdots & z_p^1 \\ z_1^2 & z_2^2 & \cdots & z_p^2 \\ \vdots & \vdots & \vdots & \vdots \\ z_1^{N-1} & z_2^{N-1} & \cdots & z_p^{N-1} \end{bmatrix} \begin{bmatrix} B_1 \\ B_2 \\ \vdots \\ B_p \end{bmatrix} \tag{2.38}$$

Note that these equations are nonlinear in z and that we need $N \geq 2p + 1$.

The method to solve the above equations may be described as follows: Let z_i for $i = 1, 2, \ldots, p$ be the roots of some p_{th} order polynomial $\Pi(z)$:

$$\Pi(z) = (z - z_1)(z - z_2)\ldots(z - z_p) = z^p - \alpha_1 z^{p-1} - \cdots - \alpha_{p-1} z - \alpha_p = 0 \tag{2.39}$$

Now, multiply the first equation in (2.37) by $-\alpha_p$, the second equation by $-\alpha_{p-1}\ldots$, the pth equation by $-\alpha_1$, and the $(p+1)$th equation by 1, and add the results. Then,

$$\begin{aligned}
y(p) &- \alpha_1 y(p-1) - \alpha_2 y(p-2) - \cdots - \alpha_p y(0) \\
&= B_1(z_1^p - \alpha_1 z_1^{p-1} - \alpha_2 z_1^{p-2} - \cdots - \alpha_p z_1^0) + B_2(z_2^p - \alpha_1 z_2^{p-1} - \alpha_2 z_2^{p-2} - \cdots - \alpha_p z_2^0) + \cdots
\end{aligned} \tag{2.40}$$

Since (2.39) is true for all z_i, the right-hand members of the above equation vanish and it is obtained that

$$y(p) = \alpha_1 y(p-1) + \alpha_2 y(p-2) + \cdots + \alpha_p y(0) \tag{2.41}$$

Applying repeatedly the above process starting this time from the second equation in (2.37), and next from the third equation, and so on, the following $N - 1 - p$ set of linear equations can be obtained:

$$
\begin{bmatrix} y(p) \\ y(p+1) \\ y(p+2) \\ \vdots \\ y(N-1) \end{bmatrix} = \begin{bmatrix} y(p-1) & y(p-2) & \cdots & y(0) \\ y(p) & y(p-1) & \cdots & y(1) \\ y(p+1) & y(p) & \cdots & y(2) \\ \vdots & \vdots & \vdots & \vdots \\ y(N-2) & y(N-3) & \cdots & y(N-p-1) \end{bmatrix} \begin{bmatrix} \alpha_1 \\ \alpha_2 \\ \vdots \\ \alpha_p \end{bmatrix} \tag{2.42}
$$

This set of equations can be solved directly if $N = 2p + 1$, or approximately by least squares if $N > 2p + 1$.

Once the α's are obtained, the z_i's values could be found from (2.39). Then, (2.38) become linear with known coefficients and the B_i's may be determined. Note that B_i's have the information about amplitude and initial phase of the modes, while the z_i's content the values for σ_i and ω_i of the modes.

2.3.3 Wavelet Transform and Its Improvements

The wavelet analysis has been introduced as a windowing technique with variable-sized regions to overcome the drawback of the fixed size window [15]. It has been one of the most important and fastest evolving signal processing tools of the last twenty years. Wavelet decomposition introduces the notion of scale as an alternative to frequency and maps a signal into a timescale plane which is equivalent to the time–frequency plane used in the short time FFT. Each scale in the timescale plane corresponds to a certain range of frequencies in the time–frequency plane.

The wavelet $\psi(t)$ (or mother wavelet) is a function of zero average [16]

$$
\int_{-\infty}^{\infty} \psi(t)\mathrm{d}t = 0 \tag{2.43}
$$

which is dilated with a scale parameter a, and translated by b to produce the daughter wavelets:

$$
\psi_{a,b}(t) = \frac{1}{\sqrt{a}} \psi\left(\frac{t-b}{a}\right) \tag{2.44}
$$

Different types of wavelets are often grouped into wavelet families according to their properties. The typical wavelet families are included in the Matlab® Wavelet Toolbox™ version 4.1.

The wavelet transform of $y(t)$ at the scale a and position b is computed by correlating $y(t)$ with a wavelet function ψ:

$$Wy(a,b) = C(a,b) = \int\limits_{-\infty}^{\infty} y(t)\frac{1}{\sqrt{a}}\psi^*\left(\frac{t-b}{a}\right)dt \qquad (2.45)$$

where $C(a,b)$ is the wavelet coefficient (approximately directly proportional to the amplitude of a specific mode) with the scale a (inversely proportional to the wavelet center frequency) and position b, ψ^* is the complex conjugated wavelet function. $y(t)$ denotes the original signal in continuous time t, and it can be any signal in a power system which contains the studied oscillation information. Those signals include, for example, the interconnecting line power flow between oscillating systems, the angular speed of an oscillating generator, and the voltage angle difference between oscillating systems.

The center frequency of the wavelet function f_c is defined as the frequency that maximizes the FFT of the wavelet function ψ:

$$f_c = \{f \,|\, \max \hat{\psi}(f) = \hat{\psi}(f_c)\} \qquad (2.46)$$

where f is frequency and $\hat{\psi}$ is the FFT of the wavelet function.

The wavelet center frequency changes accordingly when the wavelet scale is changed. If the scale is increased (the wavelet function is stretched), the center frequency of the wavelet function is decreased and vice versa. For example, if the wavelet function is scaled to have a center frequency of 0.3 Hz, it is referred to as "0.3 Hz wavelet."

The wavelet transform can be classified as continuous or discrete wavelet transform (DWT). In general, continuous wavelets are better for time–frequency analysis and discrete wavelets are more suitable for decomposition and compression [16].

Continuous wavelet transform (CWT) is calculated that both the parameters a and b are continuous variables in the interval of interest. In practice, both a and b are discrete when sampled data is analyzed. The wavelet transform is still referred to as continuous if a is an arbitrarily selected set of scales (selected according to frequency band in question and the needed resolution in frequency) and b is set by the signal sampling interval.

It is possible to use the DWT to reduce the amount of data produced by the wavelet transform that uses a certain subset of scales a, and positions b. The signal reconstruction will be as accurate as using CWT.

Because it can (with certain accuracy) extract the amplitudes of various frequency components of the signal along the time axis, the wavelet transform can be used in damping estimation. After that, the damping of the frequency components (modes) can be identified. The accuracy of the wavelet transform to measure time–

frequency variations of spectral components is limited by Heisenberg's uncertainty principle.

Both wavelet characteristics, time and frequency resolutions, affect the damping estimation of the mode after the wavelet transform is performed. Therefore, the mother wavelets with good enough frequency resolution are significantly to separate the modes from each other but having as good as possible time resolution also.

While damping the electromechanical oscillations, the analyzed signals always have a certain amount of measurement noise which is due to the measurement errors of the analyzed quantities (voltages, currents, and frequencies). The measurement errors are mainly caused by the inaccurate measurement of the PMUs and transformers. The ability of different mother wavelets to separate the noise from the actual mode is different but generally the wavelet transform is quite insensitive to noise in the analyzed signals.

2.3.4 Hilbert–Huang Transform

The Empirical Mode Decomposition (EMD) is first proposed by Dr. Huang in [17]. It is an algorithm for the analysis of multi-component signals that works by breaking them down into a number of amplitude- and frequency-modulated (AM/FM) zero-mean signals. In contrast to conventional decomposition methods, which perform the analysis by projecting the signal into a number of predefined basis vectors, EMD expresses the signal as an expansion of basic functions which are estimated by an iterative procedure called sifting. The essence of EMD is to identify the IMF by their characteristics in timescales in the data empirically. The principle steps of EMD are introduced as follows [17, 18]:

1. Find all the local minima and maxima of the signal $x(t)$;
2. Set up the upper envelope $u_k(t)$ and lower envelope $l_k(t)$ by the B-spline interpolation;
3. Calculate the mean value $m_k(t) = (u_k(t) + l_k(t))/2$;
4. Subtract the $m_k(t)$ from original signal and get the residue $h_k(t) = x_k(t) - m_k(t)$;
5. Repeat the steps from 1 to 4, if the $h_k(t)$ satisfies the conditions of IMF, then extract the $h_k(t)$ and record it as $c_k(t) = h_k(t)$;
6. Determine the IMFs and obtain the residue $r_n(t) = x_n(t) - c_k(t)$;
7. Check whether $r_n(t)$ is a mono-component or not; if not, then return to Step 1 and repeat the iteration process until the $r_n(t)$ only has three extrema.

The flowchart of EMD is displayed in Fig. 2.10.

In Fig. 2.10, there are two loops in the process of EMD. The internal loop is called sifting, which is used to extract the intrinsic mode functions (IMFs); and the external loop is the main iteration, which is implemented to define the numbers of

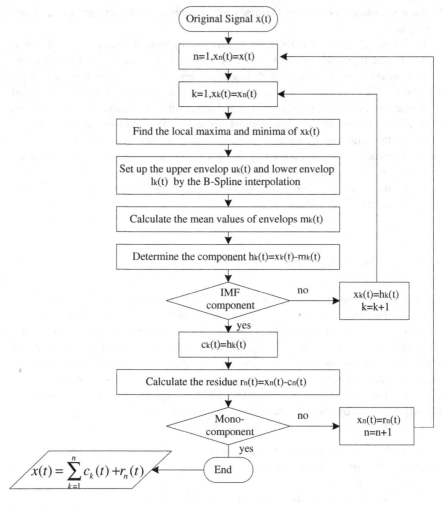

Fig. 2.10 The flowchart of EMD

IMFs and end the process of decomposition. After EMD, the signal can be expressed into two parts [19]:

$$x(t) = \sum_{k=1}^{n} c_k(t) + r_n(t) \tag{2.47}$$

$c_k(t)$ stands for the IMF and n is the numbers of IMF; $r(t)$ indicates the residue of the signal.

From the perspective of sifting process, all the IMFs should be locally orthogonal to each other because IMF is got by subtracting its local mean from the signal.

Obviously, the results are not fully orthogonal because the local mean is not the true value while calculated by the lower and upper envelopes. In [17], the index of orthogonal (IO) and energy leakage are employed to evaluate the orthogonal degree of the EMD. Empirically, in most cases in power system, the energy leakage is small and the IMFs obtained by EMD can be considered near orthogonal to each other.

From the form of analytical function, the conceptions of instantaneous amplitude and envelope of signal have been well accepted. However, the notion of instantaneous frequency is still highly controversial. In [17], it is noticed that there are still considerable defects in defining the instantaneous frequency as follows:

$$f(t) = \frac{1}{2\pi} \frac{d\varphi(t)}{dt} \qquad (2.48)$$

In (2.48), it is clear that the instantaneous frequency is a single value function of time. The instantaneous frequency is determined by the specific time. Therefore, it can only represent one component named "mono-component." But it is difficult to judge whether a function is "mono-component" or not because there is no precise and strict definition.

From the description above, notion of instantaneous frequency is based on the assumption that there is only one single frequency component at each time instant. Thus, under the condition that $x(t)$ is a multi-component signal, the obtained results are meaningless. A set of synthetic signals is used to illustrate this problem:

$$\begin{cases} x_1 = \sin(1.2\pi t) \\ x_2 = 1.5e^{-0.1t} \sin(1.5\pi t) \\ x_3 = x_1 + x_2 \end{cases} \qquad (2.49)$$

It is clear that x_1 is a pure sine signal, x_2 is an exponential decay signal, and x_3 is a composite signal. At the given time t, x_1 and x_2 have the unique frequency. The value has no any relations with the values at other times. However, the instantaneous frequency of x_3 is meaningless because it is a superposition value. The comparison results are displayed in Fig. 2.11.

Figure 2.11 compares the analytical functions, instantaneous amplitudes, and instantaneous frequencies of three original signals. The main conclusions can be summarized as follows: (1) for the analytical functions, x_1 is a standard circle while there are obvious distortions on the both ends of x_1 and x_2; (2) the instantaneous frequency of x_1 is equivalent to 0.6 Hz at all the time range, the instantaneous amplitude is one constant; (3) for x_2, its frequency is nearly to 0.75 Hz except the both end areas of the signal, these phenomena can also be seen from the instantaneous amplitudes, and it will be discussed in next chapter; and (4) both the amplitude and frequency of x_3 are irregular and insignificant.

In order to get the accurate instantaneous frequency, a new definition of a class of function named intrinsic mode function (IMF) is proposed in [17]. The IMF should satisfy two conditions: (1) the number of extrema and the number of zero crossings must either equal or differ at most by one in the whole data range; (2) at

(a)

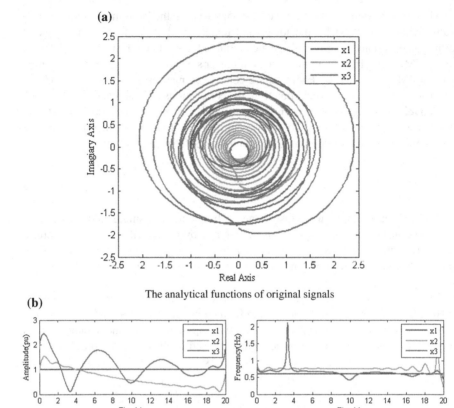

The analytical functions of original signals

(b)

The calculated instantaneous amplitudes and frequencies

Fig. 2.11 The comparison of instantaneous parameters

any point, the mean value of the upper envelope defined by the local maxima and the lower envelope defined by the local minima is zero.

From the definition of IMF, it is clear that the IMF is adopted because it represents the oscillation mode imbedded in the data. In order to define the instantaneous frequency clearly, the EMD is employed to decompose the single nonlinear signal into the needed IMFs.

Hilbert–Huang Transform

In this section, the analytical function is introduced first. In general, the HT form of a real function $x(t)$ is defined as follows [17, 20]:

$$y(t) = \frac{1}{\pi} P \int_{-\infty}^{+\infty} \frac{x(\tau)}{t - \tau} d\tau \tag{2.50}$$

where P indicates the Cauchy principal value.

Clearly, the HT form of $x(t)$ is a linear operation. The inverse HT, by means of the original signal $x(t)$ recovered from $y(t)$, is calculated by

$$x(t) = -\frac{1}{\pi} P \int_{-\infty}^{+\infty} \frac{y(\tau)}{t - \tau} d\tau \tag{2.51}$$

The detailed introduction of HT with the emphasis on its mathematical foundation can be found in [20]. $y(t)$ is interpreted as the convolution of $x(t)$ with the time function $1/\pi t$. Therefore, it gives the emphasis on the local properties of $x(t)$. It is well known that the convolution of two functions in time domain can be transformed into the multiplication of their FFTs in frequency domain. Therefore, HT can also be interpreted as a $\pi/2$ phase shifter, which means the amplitudes of all frequency components in the signal are unchanged while the phases of them are shifted by $\pi/2$.

Comparing the FFT, HT has a number of useful properties for analyzing the LFO signal [20]: (1) $x(t)$ and $y(t)$ have the same magnitude spectrum; (2) the HT form of $y(t)$ is $x(t)$; (3) $x(t)$ and $y(t)$ are orthogonal over the entire time interval; and (4) HT form of $\dot{x}(t)$ is equivalent to the derivative of $y(t)$.

According to the relationship between $x(t)$ and $y(t)$, the form of analytical function is defined as follows:

$$z(t) = x(t) + jy(t) = A(t)e^{j\phi(t)} \tag{2.52}$$

$$\begin{cases} A(t) = \sqrt{x^2(t) + y^2(t)} \\ \phi(t) = \arctan(y(t)/x(t)) \end{cases} \tag{2.53}$$

where $A(t)$ is the instantaneous amplitude, and $\phi(t)$ is the instantaneous phase.

The analytical signal represents a time-dependent phasor in the complex plane with instantaneous amplitude and phase. Figure 2.12 gives two conceptual representations of the analytical signal.

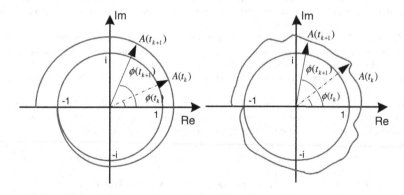

Fig. 2.12 Two conceptual representations of the analytical signal

Theoretically, although many methods have been proposed to define the imaginary parts of the analytical function, the HT is the unique and convenient way to set up the analytical function.

2.4 Summary

This chapter presents the theoretical foundation of LFO. Two techniques are presented to analyze the LFOs, one based on system model and the other based on measured information. Model-based method has been widely utilized to obtain the model of power system to design the controller. The key idea of model-based methods is that the whole system is linearized around a specific operating point. Then the system model is described as the state equations. The eigenvalues, eigenvectors, and participation factors of the state matrix are calculated. The oscillatory parameters including the frequency and damping ratio of each mode are determined by the eigenvalues. While the measurement-based methods has been investigated for extracting model information from measured system responses, most of the methods based on the ensemble measurement matrix can extract the oscillation mode shape or similar results. However, they cannot provide time-varying and nonlinear oscillatory parameters of each single measured signal. Furthermore, the calculated results cannot be directly utilized to estimate the stability of the oscillation mode.

References

1. Rogers G (2000) Power system oscillations. Kluwer Academic Publishers, Norwell. ISBN 0-7923-7712-5
2. DeMello FP, Corcordia C (1969) Concept of synchronous machine stability as affected by excitation control. IEEE Trans Power Appar Syst 88(4):316–329
3. Xu YH, ZH JB (2011) The resonance mechanism low frequency oscillations induced by nonlinear governor system. In: International conference on business management and electronic information (BMEI), 13–15 May 2011
4. Yu YP, Min Y, Chen L et al (2011) The disturbance source identification of forced power system oscillation caused by continuous cyclical load. In: International conference on electric utility deregulation and restructuring and power technologies (DRPT), 6–9 July 2011
5. Seydel R (2009) Practical bifurcation and stability analysis, 3rd edn. Springer, New York. ISBN 978-1-4419-1739-3
6. Wen XY (2005) A novel approach for identification and tracing of oscillatory stability and damping ratio margin boundaries. A Dissertation for PhD degree, Iowa State University
7. Prabha K (2004) Power system stability and control. McGraw-Hill, New York
8. Rogers G (2000) Power system oscillations. Kluwer
9. Klein M, Rogers GJ, Kundur P (1991) A fundamental study of inter-area oscillations. IEEE Trans. Power Syst 6:914–921
10. Kundur P (1993) Power system stability and control. McGraw-Hill, New York

11. Sadikovi'c R (2006) Use of FACTS devices for power flow control and damping of oscillations in power systems. PhD Dissertation, Swiss Federal Institute of Technology Zurich
12. Phadke AG, Thorp JS (2008) Synchronized phasor measurements and their applications. Springer, New York
13. Restrepo JQ (2005) A real-time wide-area control for mitigating small-signal instability in large electric power systems. A Dissertation for PhD degree, Washington State University
14. Trudnowski DJ, Johnson JM, Hauer JF (1998) SIMO system identification from measured ringdowns. In 1998 Proceedings of the American control conference, pp 2968–2972
15. Zemmour AI (2006) The Hilbert-Huang transform for damage detection in plate structures. A Dissertation for PhD degree, Master of Science
16. Turunen J (2011) A wavelet-based method for estimating damping in power systems. Doctoral Dissertations, Aalto University Publication Series
17. Huang NE, Shen Z, Long SR et al (1998) The empirical mode decomposition and the Hilbert spectrum for nonlinear and non-stationary time series analysis. Proc R Soc Lond A 454:903–995
18. Messina AR (2009) Inter-area oscillations in power systems: a nonlinear and non-stationary perspective. Springer, Berlin. ISBN 978-0-387-89529-1
19. Yuan ZY, Tao X, Zhang Y C et al (2010) Inter-area oscillation analysis using wide area voltage angle measurements from FNET. IEEE power and energy society general meeting 25–29 July 2010
20. Li W, Robert MG, Dong J Y et al (2009) Wide area synchronized measurements and inter-area oscillation study. IEEE/PES power systems conference and exposition (PSCE), 15–18 Mar 2009

Chapter 3
Oscillatory Parameters Computation Based on Improved HHT

In the point of operation, the responses of real bulk power system are dynamic and time-varying owing to the influences of load variations, topology changes and control actions. It is necessary to monitor and calculate the instantaneous oscillatory parameters so as to provide an optimal control strategy. As a nonlinear and non-stationary signal processing approach, Hilbert–Huang transform (HHT) which contains Empirical Mode Decomposition (EMD) and Hilbert Transform (HT) has been widely utilized to calculate the amplitudes and frequencies of the single LFO signal. The decomposition ability of EMD provides the condition for HT and HT extends the application fields of EMD. However, the further applications in power system are limited by the inherent shortcomings of HHT, such as the end effects (EEs) and mode-mixing in the process of EMD as well as Gibbs phenomena in traditional HT. In this chapter, the shortcomings of HHT are analyzed and an improved HHT is proposed to determine the oscillatory parameters of single measured signal.

3.1 Introduction of Improved Empirical Mode Decomposition (EMD)

3.1.1 The Selection of Stop Criterion for Sifting in EMD

It has been proven that the HHT can trace and extract the dynamic characteristics of the signal. However, it has some inherent shortcomings because there is no mathematical foundation in the sifting process. From the steps of EMD, it is clear that the sifting stop criterion has direct relationship with the decomposition accuracy and decomposition speed. Generally speaking, three parameters, including amplitude, frequency, and phase, are considered as the sifting stop criteria. Aiming at the application of HHT in analysis of the LFO signal, the existing sifting stop criteria are evaluated and the results are displayed in Table 3.1.

By comparison in Table 3.1, the sifting stop criterion based on the three parameters has a good performance in decomposition accuracy and computation

© Springer-Verlag Berlin Heidelberg 2016
Y. Li et al., *Interconnected Power Systems*, Power Systems,
DOI 10.1007/978-3-662-48627-6_3

Table 3.1 Comparisons of different sifting stop criteria

Name	Formula	Default value	Characteristic	Orthogonality	Calculated time
Standard deviation (SD)	$SD = \sum\limits_{t=0}^{T} \left[\dfrac{\left|h_{(k-1)}(t) - h_k(t)\right|^2}{h_{(k-1)}^2(t)} \right]$	SD = 0.3	Standard deviation	good	common
Three parameters	$\begin{cases} m(t) = \dfrac{|u_k(t) + l_k(t)|}{2} \\ a(t) = \dfrac{|u_k(t) - l_k(t)|}{2} \\ \sigma(t) = |m(t)/a(t)| \\ \sigma(t) < \theta_1(1 - \alpha); \sigma(t) < \theta_2(\alpha) \end{cases}$	$\theta_1 = 0.05$ $\theta_2 = 10$ $\theta_1 = 0.5$ $\alpha = 0.05$	Instantaneous features	good	good
S-number	****	$k = 2000$	Define the sifting number	good	flexible
Frequency range	$B^2 = B_a^2 + B_f^2$ $B_a^2 = \int (a'(t)/a(t))^2\, a^2(t)\,\mathrm{d}t$ $B_f^2 = \int (\phi'(t) - \langle\omega\rangle)^2\, a^2(t)\,\mathrm{d}t$	****	Mono-frequency component	common	common

time. Furthermore, the sifting ability of three parameter criterion can be adjusted by changing the values of θ_1, θ_2, and α aiming at different signals. Therefore, the three parameters stop criterion is selected as the default condition in the process of EMD [1–4].

3.1.2 End Effects and Extrema Symmetrical Extension

When the HHT is employed to analyze the LFO signal, there are two kinds of EEs: one is produced in the HT and the other is in the process of EMD. The former involves in the limitations of Bedrosian and Nuttall theorems and it will be analyzed in the following section. The latter is much harder to be resolved because it involves in the B-spline fitting function [1].

During the decomposition process, B-spline function is used to construct upper and lower envelopes of data in order to determine the mean values in the iterative sifting processes until satisfying the stop sifting criterion. Both end areas of the signal appear divergence problems because the interpolating points of the B-spline function at the ends are not decided. The divergences influence the inside data gradually with the sifting procedures and it leads to fake components and an overestimated energy of the signal. Actually, the EEs have been becoming an intrinsic shortcoming of EMD.

In order to solve EEs in process of EMD, many extension methods have been proposed in the existing publications. All the extension methods can be classified into two types: one is based on the end points and extrema, and the other is based on the intelligent algorithm. In most cases, the intelligent extension methods can reduce the EEs considerably, but the extending speed is so slow that it can only be used in offline signal processing. In this section, four kinds of simple extension methods, named Envelop Linear Extension (ELE) [5], Envelop Extrema Extension (EEE) [6], Extrema Symmetrical Extension (ESE) [7], and Local Wave-matching Extension (LWE) [8], are briefly introduced. All of these methods are based on the characteristics of the end points and extrema. The main steps of these methods are summarized as follows:

(1) Judge the characteristics of first and end point by comparing with the first and last extrema, respectively;
(2) Extend the signal according to the characteristics of beginning and ending segments;
(3) Determine all extrema of the extended signal;
(4) Interpolate the extended signal and construct the upper and lower envelopes;
(5) Calculate the mean values of the upper and lower envelopes;
(6) Intercept the parts of original signal.

In this section, four mentioned extension methods are compared using a test signal in order to decide which one is most suitable for the LFO signal analysis. The test signal is expressed as follows:

$$x = x_1 + x_2 = 10e^{0.1t}\cos(2.4\pi t) + 10e^{-0.1t}\cos(\pi t) \qquad (3.1)$$

Both x_1 and x_2 satisfy the definition of the intrinsic mode function (IMF) [1] and they are considered as two IMFs. In order to evaluate the performances of these extension methods, four criteria are introduced as follows: left residue, decomposition time, means of IMF1 in first sifting process, and fitting degrees of IMFs.

The left residue is defined as the left signal extracted from the original signal by subtracting the IMF1 and IMF2. Ideally, the residue should be zero because there is no constant or trend component in the test signal. If the residue is bigger, the decomposing precision is unreliable. On the other hand, the essential of EEs is that the lower and upper envelopes of the signal are deformed. Under this condition, the EMD produces some fake components in the process of decomposition. The error of the left residue is set as

$$\text{re\_err} = \sqrt{\sum_{i=1}^{n} r^2(i) \Big/ n} \qquad (3.2)$$

where $r(i)$ stands for the left residue sequence; n means the number of the sampling points.

The residues calculated by the proposed four different extension techniques are shown in Fig. 3.1a and Table 3.2.

Fig. 3.1 The comparisons of four extending approaches (a) the left residues (b) the means in the IMFs (the first sifting)

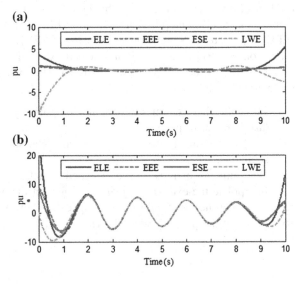

Table 3.2 Comparisons of four extension techniques

Name	ELE	EEE	ESE	LWE
Decomposition time (s)	0.0456	0.0384	0.0382	0.1124
re_err	0.1676	0.0124	0.0158	0.0893
IMF1_err	0.0130	0.0082	0.0064	0.0163
IMF2_err	0.0309	0.0176	0.0221	0.0170
Number of IMFs	3	3	3	4

It is clear that the mean values of upper and lower envelopes in the sifting steps are essential for the accuracy of determination of the IMFs. The most important reason for the appearance of the EEs is that the upper and lower envelopes at the ends are divergent. With the decomposition of the EMD, the influences of the EEs become stronger and stronger. As for the test signal, the first mean values in sifting process in IMF1 which are calculated by the above four extension techniques are compared in Fig. 3.1b.

Calculation time is also a very important indicator to evaluate the ability of the proposed extension techniques. For the EMD, the decomposing time has direct relationship with the estimation of oscillatory parameters of IMF. The decomposition time of EMD with four different extension methods in the same personal computer are calculated and displayed in Table 3.2, respectively.

The upper and lower envelopes in the sifting process swing at the two ends of the data sequence. The cause of the swing is the uncertainty of the interpolating point of the B-spline function at the end. The result of this swing gradually pollutes inside the part of signal or even the whole data sequence. Consequently, it makes the result seriously distortions. According to the decomposition processes and stop sifting criterion, comparison of each value of IMF and the corresponding component in the original signal can be considered as an evaluation indicator. This indictor is expressed as

$$\text{IMF}_k\_\text{err} = \sqrt{\frac{\sum_{i=1}^{n} \left(x_k^2(i) - \text{IMF}_k^2(i) \right) \Big/ \sum_{i=1}^{n} \left(x_k^2(i) \right)}{n}} \tag{3.3}$$

where k is the number of original signal or corresponding IMF. For the test signal, the decomposition error of IMFs is calculated and shown in Fig. 3.2.

According to the computation results, some conclusions are summarized as follows: (1) the computation times of EMD with EEE and ESE are nearly the same. The LWE takes more time than other three approaches because LWE needs to calculate the similarity of extrema which are nearest to the ends and other extrema; (2) for the error of residue, EEE has the best decomposition results. The left residue obtained by LWE contains the biggest error; (3) for the mean values shown in

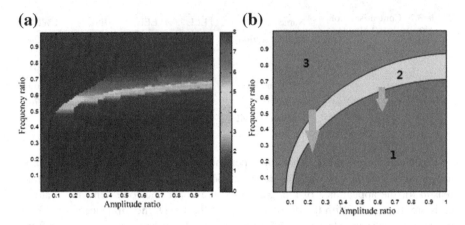

Fig. 3.2 The interpretation of mode-mixing under different amplitude/frequency ratios (**a**) the decomposition results (**b**) the simplified interpretation

Fig. 3.1b, the middle parts of IMF1 which are obtained from the four extension methods are nearly the same. However, there are different convergences at the ends. EEE and ESE have better performance than ELE and LWE; (4) the errors between IMF1 and x_1 are smaller than the errors between IMF2 and x_2. The reason is that the error in IMF1 affects the decomposition accuracy of following IMF2; (5) because of EEs, it produces the fake component in the process of EMD; (6) comparing with other three techniques, ESE has better performance not only in the decomposition time but also in the decomposition accuracy. Therefore, ESE is added into the sifting process of EMD to analyze the LFO signal.

It should be noted that it is incomplete and inaccurate to evaluate the performances of the different extension approaches by test signal. However, the proposed synthetic signal stands for the typical and common oscillation mode in power system and the proposed evaluation criteria contain all the respects of IMF. Thus, the comparative results are credible and convincible.

3.1.3 Mode-Mixing and Frequency Heterodyne Technique (FHT)

Although EMD principle is clear and appealing, the performance analysis is difficult because there is no mathematical or theoretical foundation. Furthermore, there are some shortcomings in the sifting process. In [9] and [10], they have shown that the decomposition results are incorrect when the parameters of original components have specific relationships. In LFO analysis, the mode-mixing is defined as the specific oscillation mode is decomposed into two IMFs in time window. Supposing that there are two original components in the original signal:

$$x = x_1 + x_2 = A_1 \sin(2\pi f_1 t) + A_2 \sin(2\pi f_2 t) \tag{3.4}$$

x_1 is the higher frequency component and x_2 is the lower frequency component. The decomposition results under different amplitude ratios and frequency ratios are shown in Fig. 3.2. Here, the evaluation is set as the ratio between IMF1_err and the higher frequency component. If there is mode-mixing phenomenon in the process of EMD, IMF1 would contain the lower frequency component. The frequency ratio is defined as f_2/f_1 and the amplitude ratio is defined as A_2/A_1.

From Fig. 3.2 and the theory in [11], there are three areas in the plane: (1) in Area 1, the relationships between amplitude ratio and frequency ratio can be expressed approximately as $A_2/A_1 \leq 2.4(f_1/f_2)^{1.75}$. The original components can be separated by EMD and the decomposition error is small; (2) in Area 2, although the original components can be extracted by several decomposition times, the decomposition results are incorrect. The relationship between amplitude ratio and frequency ratio can be defined as $(f_1/f_2)^2 \leq A_2/A_1 \leq 2.4(f_1/f_2)^{1.75}$; (3) Area 3 indicates that the original components are considered as one IMF in the process of EMD. In this area, the relationship between amplitude ratio and frequency ratio is $A_2/A_1 \leq (f_1/f_2)^2$ or $f_2/f_1 \geq 0.66$. Therefore, it is clear that the key aspect to avoid the mode-mixing phenomenon in EMD is to change the amplitude and frequency relationships from Area 3 or Area 2 to Area 1.

Aiming at the mode-mixing phenomenon in EMD, many methods have been proposed. These methods can be classified into two groups: (1) add the masking signals in the sifting process [9, 12]; (2) combine EMD with other signal analysis methods [13]. It is really difficult to appraise their performances because each improved method has its own characteristics and application fields. In communication system, a shift of the range of frequencies in a signal is accomplished using linear modulation, which is defined as the process by which some characteristics of a carrier is varied in accordance with a modulating wave. From this point, the FHT is used to modulate the original signal in the process of EMD and to reduce the influence of mode-mixing. Generally speaking, the measured signal from phasor measurement unit (PMU) contains several oscillation modes and the characteristic of each mode is nearly to the constant in specified time. Furthermore, the local mode and inter-area mode have different existing times. These features provide good applying conditions for FHT [10, 14].

The complex representation of the analytical function is $\tilde{s}(t) = s(t) + j s_H(t)$, and $s_H(t)$ is the HT form of $s(t)$. It is in accordance with the form of single-sideband modulation (SSB). With the theory of modulation, the upper sideband signal is obtained as (3.5)

$$f_{\text{USSB}}(t) = \text{Re}\left\{(s(t)+js_{\text{H}}(t))e^{j2\pi Ft}\right\} \qquad (3.5)$$

Similarly, the lower sideband signal is obtained as follows:

$$f_{\text{LSSB}}(t) = \text{Re}\left\{(s(t)+js_{\text{H}}(t))e^{-j2\pi Ft}\right\} \qquad (3.6)$$

The analytical signal $\tilde{s}(t)$ may be expressed as a time-varying phasor position at the original of the (p,q)-plane, as indicated in Fig. 3.3a. Figure 3.3b shows the phasor representation of the complex exponential $e^{j2\pi Ft}$. By the definition in (3.6), the analytical signal $\tilde{s}(t)$ is multiplied by the complex exponential $e^{j2\pi Ft}$. The angles of these two phasors are added and their lengths are multiplied, as shown in Fig. 3.3c.

The principle of the lower sideband is the same as the upper sideband [10]. The application of FHT in the EMD is explained in Fig. 3.4.

In Fig. 3.4, it is clear that the application of the FHT in EMD can be explained from three steps: (1) original signal is transferred into the shifted signal by multiplying the shift factor; (2) the shifted signal is decomposed into shifted IMFs and residue; and (3) the shifted IMFs and residue are restored by the reductive factor.

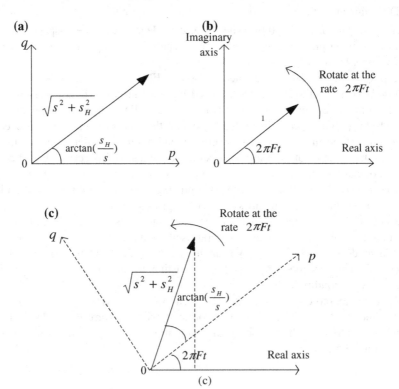

Fig. 3.3 Modulation principle of upper signal

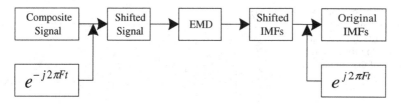

Fig. 3.4 Application of FHT in EMD

It is clear that the results of EMD can be modified by changing of the frequencies in the multifrequency signal. Therefore, the choice of shifting frequency is an essential part for the application of FHT.

Supposing that there are two components in the composite signal:

$$x = x_1 + x_2 = \Lambda_1 e^{\sigma_1 t} \cos(2\pi f_1 t) + \Lambda_2 e^{\sigma_2 t} \cos(2\pi f_2 t) \qquad (3.7)$$

The shifting frequency factor is chosen as

$$F = 2f_1 - f_2 \qquad (3.8)$$

The frequencies in the shifted signal are calculated as follows:

$$\begin{cases} f_{11} = F - f_1 = 2f_1 - f_2 - f_1 = f_1 - f_2 \\ f_{12} = F - f_2 = 2(f_1 - f_2) = 2f_{11} \end{cases} \qquad (3.9)$$

The performances of the FHT are tested by testing signal.

$$x(t) = x_1(t) + x_2(t) = 10e^{-0.1t} \cos(1.6\pi t) + 12e^{-0.05t} \cos(\pi t) \qquad (3.10)$$

Here, the sampling frequency is 100 Hz and the length of time window is 20 s. It is obvious that the amplitudes of two original components are time-varying and the relationships between amplitudes and frequencies are also time-varying. The original signal and its time–frequency spectrum are shown in Fig. 3.5.

According to the results in Fig. 3.2 and the theory in [11], the time when mode-mixing takes place can be calculated. In order to demonstrate the validity and accuracy of FHT, the frequency distributions of original signal $x(t)$, shifted signal $xs(t)$, and reductive signal $xr(t)$ are compared in Fig. 3.6.

The time when the mode-mixing occurs is nearly equal to 6.58 s. Here, the frequency of shifting factor is selected as 0.9 Hz, and the frequencies of original components are changed into 0.1 (0.9–0.8) and 0.4 (0.9–0.5) Hz. Under these conditions, the complex signal is moved into Area 1 from Area 2, and the original components can be separated clearly, and then the shifted signals are decomposed by EMD. Reductive IMFs are achieved by reverse algorithm. The comparisons of original components, original IMFs, and reductive IMFs are displayed in Fig. 3.7

Fig. 3.5 The original signal and its time–frequency spectrum (**a**) the original signal (**b**) the time–frequency spectrum of the original signal

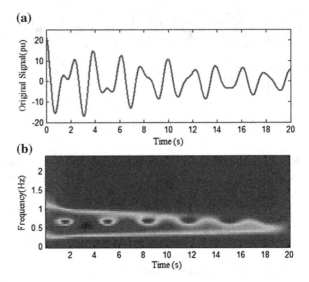

Fig. 3.6 Fast Fourier transform (FFT) [14] spectrums of original signal, shifted signal, and reductive signal

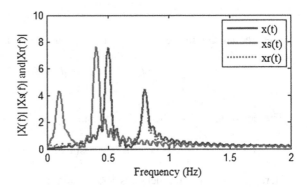

In Fig. 3.7, the IMF1 directly decomposed by EMD is expressed as the combination of $x_1(t)$ and $x_2(t)$ after 6.58 s. It is clear that there are obvious mode-mixing phenomena in IMF1 and IMF2. However, the decomposition results obtained by the EMD with FHT have good fitting with original components. Therefore, the proposed FHT can reduce the influence of mode-mixing effectively.

In order to display the mode-mixing both in time and frequency domains, the time–frequency spectrums of original IMFs and reductive IMFs are compared in Fig. 3.8.

From the results in Fig. 3.8, the main conclusions can be summarized as follows: (1) For the EMD, both IMF1 and IMF2 have the mode-mixing phenomena. IMF1 contains different instantaneous frequencies and irregular amplitudes with time. The instantaneous frequencies of IMF2 are near to 0.2 Hz in 7–20 s. This means that the mode-mixing in IMF1 affects the accuracy of IMF2. Actually, it is the main obstacle for the further application of HHT in analysis of measured LFO signal.

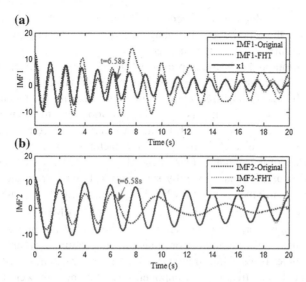

Fig. 3.7 Comparisons of original components, original IMFs, and reductive IMFs

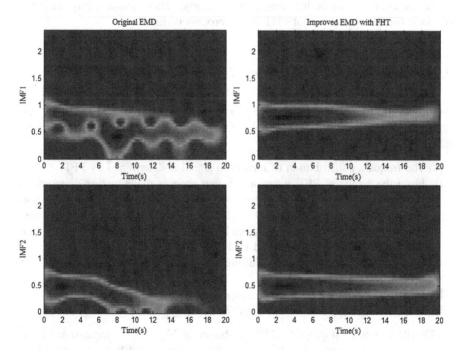

Fig. 3.8 The time–frequency spectrums of IMFs

(2) IMF1 and IMF2 which are obtained by the improved EMD are mono-frequency components and the amplitudes are changing regularly. The improved EMD can avoid the mode-mixing phenomenon effectively.

As the description above, the improved EMD with FHT can overcome special mode-mixing problem, but the shifting frequency factor has direct relation with its performance. If the shifting frequency is too big, it is difficult to guarantee the ability of dealing with the mode-mixing problem; if too small, the frequencies in the shifted signal are near to zero, which results in increasing the errors of EMD and energy leakage. The choice of shifting frequency factor does not only depend on the initial parameters of the signal, but also depends on the dynamics characteristics of the time-varying oscillations. Generally speaking, just considering the frequencies parameters in multifrequency signal, when the shifting frequency is in the range of $f_1 + \frac{f_1 - f_2}{2} < F < 2f_1 - f_2$, the decomposing accuracy of EMD is receivable and credible. It should be noted that there are two conditions for the application of FHT: one is the frequency distributions of original components in the complex measured signal and the other is the higher frequency components which have faster decaying speed than the lower frequency components. In the bulk power systems, the complicated dynamic characteristics are usually determined by one or two oscillation modes. Furthermore, the local mode has higher frequency and damping ratio, while the inter-area mode has longer existing time. These features provide good platform for the application of FHT in the process of EMD.

3.1.4 The Improved EMD Based on ESE and FHT

Aiming at the EEs and specific mode-mixing in the EMD, the ESE and FHT are added into the process of EMD. Through the analysis above, the steps of improved EMD are introduced as follows:

(1) The frequency distribution of measured signal is determined by FFT;
(2) From the results of FFT, estimate whether there is mode-mixing phenomena between first high frequency and second high frequency or not, if not, decompose the original signal by EMD based on ESE in every sifting process; if yes, suitable shifting parameter can be defined by frequency factor, and then the shifted signal is decomposed by EMD based on ESE technique;
(3) Recover IMF1 by multiplying shifting frequency factor, and then get the first residue $r_1(t) = x_1(t) - \text{IMF}_1$;
(4) Repeat Step3. Step4 until the corresponding IMFs are extracted;
(5) If $r_n(t)$ is a mono-component, the decomposing process would be ended.

The flowchart of improved EMD is shown in Fig. 3.9. Compared with the original EMD, the improved EMD reduces the influences of EEs in the process of sifting and the FHT can overcome the specific mode-mixing. The decomposition results can be divided into two parts: $c_j(t)$ stands for IMF and $r(t)$ is the mono-component in the original signal.

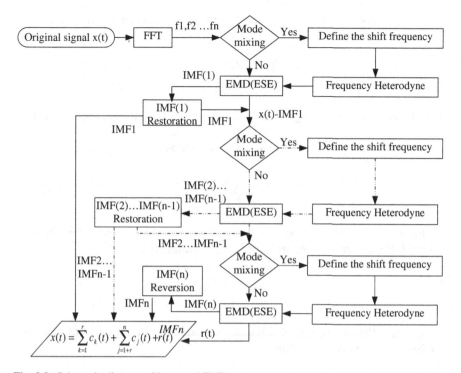

Fig. 3.9 Schematic diagram of improved EMD

3.2   Time and Frequency Analysis of Intrinsic Mode Function

According to the theory of improved EMD, the nonlinear and nonstationary signal can be decomposed into a set of IMFs locally and adaptively. 'Locally' means that the EMD can identify the variation of the instantaneous parameters; 'Adaptively' stands for the extracted IMFs with time-dependent amplitudes and phases information. In this section, how to identify the IMFs which mean the specific oscillation mode is researched based on the time characteristics and frequency distribution analysis of decomposed IMFs.

From the point of time domain, it is clear that the IMFs containing the physical meaning components have the higher energy than the higher frequency components or noises. Furthermore, Huang has pointed out that although the result of EMD cannot be proved orthogonal in theory [1], the practical decomposition result is usually near to orthogonal. Therefore, the energy of the original signal is equal to the sum of IMFs' energies and residue's energy.

In [13], the author proposed a method to calculate the energy of signal. The basic principles of this method are introduced in this part.

For a specified signal $\delta(t)$, the energy criterion is described as follows:

$$E_0 = \int_{t_1}^{t_2} \delta^2(t)\mathrm{d}t \qquad (3.11)$$

where E_0 is the energy of original signal; t_1 denotes the start time; and t_2 indicates the end time.

The discretized form of (3.11) can be expressed as

$$E_0 = \sum_{i=1}^{N} \delta^2(i) \qquad (3.12)$$

After EMD, the original signal $\delta(t)$ is decomposed into several IMFs and one residue. IMFs and residue are near orthogonal. Furthermore,

$$E_{\mathrm{sum}} = E_r + \sum_{i=1}^{n} E_{\mathrm{IMF}}(i) \qquad (3.13)$$

where E_{sum} is the total energy of IMFs and residue; E_r is the energy of the residue; and $E_{\mathrm{IMF}}(i)$ is the energy of IMF_i.

$$\varepsilon = \frac{E_n - E_0}{E_0} \times 100\% \qquad (3.14)$$

where ε is the energy decomposition error. Actually, ε can also be used to indicate the orthogonal feature of EMD, if ε equals to zero, which means that the IMFs are fully orthogonal.

$$\eta(i) = 100\% \times E_{\mathrm{IMF}}(i) \left/ \sum_{i=1}^{n} E_{\mathrm{IMF}}(i) \right. \qquad (3.15)$$

where $\eta(i)$ is the energy coefficient of IMF_i. This parameter is used to describe the energy relations among IMFs. Obviously, the energy of IMF which stands for the meaningful oscillation mode is higher.

According to the theory in [15] and [16], the mean values of IMFs contain useful information in determining the noise and physical oscillation mode. By decomposition, the number of extreme point decreases during its shifting from one residual to the next. This feature guarantees that the complete decomposition can be achieved in a finite number of steps. However, different sifting stop criteria affect the numbers of IMFs and their corresponding mean values. Here, the mean values are proposed as a supplemental criterion to calculate the number of IMFs which contain the plentiful oscillatory information.

A key property of EMD is its ability to act as a filter. Actually, the process of EMD can be considered as the separation of different frequency components. For the measured signal from actual power system, it is always polluted by many kinds of noises. First, the higher frequency components or noises are extracted and they are denoted as the fore-IMFs. Then, the meaningful components polluted by noise in original signal are identified and they are saved as the middle-IMFs. Next, the artificial components or ultra-low frequency components are left. And the last extracted function is usually mono-tone, or has one extreme. According to the above descriptions and the principles of EMD, the decomposition results of complex measured data can be divided into three parts: the noise or higher frequency components, the physical oscillation modes, and the trend.

From the perspective of frequency domain, it has been obviously proved that the range of the low frequency is 0.1–2.5 Hz. Therefore, the frequencies of IMFs located fully or partially in mentioned range can be considered as the physical and meaningful components.

From the analysis above, it is clear that the EMD provides a method to identify the original components in the measured signal. This method is also used to extract the trend parts and noise part in Chap. 4.

3.3 Normalized Hilbert Transform (NHT)

Theoretically, HT can get acceptable results only under the condition that the signals obey the Bedrosian and Nuttall theorems. In this section, the synthetic signal is used to display the limitations of HT when it is used to determinate the analytical form of IMF. In order to avoid the influence of EEs caused by EMD, the pure sinusoid and incremental exponential sinusoidal function are chosen as the testing signal.

$$\begin{cases} x_1 = A_1 \sin(2\pi f_1 t) \\ x_2 = A_2 e^{-\sigma_2 t} \cos(2\pi f_2 t + \pi/3) \end{cases} \tag{3.16}$$

where $A_1 = 6$ p.u, $A_2 = 4$ p.u; $\sigma = -0.05$ p.u; $f_1 = 0.6$ Hz, $f_2 = 0.5$ Hz. HT is employed to calculate the instantaneous amplitudes and frequencies of the x_1 and x_2, respectively. The calculated results are compared in Fig. 3.10.

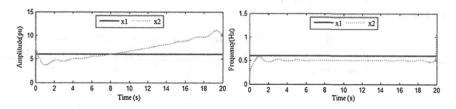

Fig. 3.10 Calculated instantaneous parameters by HT

In Fig. 3.10, it is clear that both the instantaneous amplitudes and frequencies of x_2 have obvious aberrations at the ends of signals. Comparing with the computation results of x_2, the values of x_1 are more accurate and there are no distortions at the ends. This gives us a revelation how to reduce the influence of the Gibbs phenomena in HT. Based on these hints, the normalized Hilbert transform (NHT) is proposed in [17]. However, how to use the decomposed AM and FM to explore the dynamic characteristics which contained in the IMF is not investigated sufficiently and completely. Furthermore, it just provided the method how to calculate the instantaneous frequency. As for the LFO signal from the actual power system, the measured information displays highly nonlinear and time-varying characteristics. Furthermore, the instantaneous damping ratio is the most important indicator because it has direct relationships with the security of the power system. Therefore, calculation of instantaneous damping ratio has great significance in the analysis of oscillatory signal. Here, the calculations of instantaneous parameters of IMFs which stand for the LFO modes using NHT are introduced briefly.

3.3.1 Decompose the IMF into AM and FM Parts

This subsection gives a simple description on the normalized method. The detailed information can be found in [17].

- Express the absolute value of IMF as |IMF|;
- Identify all the local maxima of the |IMF|;
- Connect the maxima points (including the two end points) with a B-spline curve and the B-spline curve $e_1(t)$ is considered as the empirical envelope of the IMF;
- Calculate the normalized data $y_1(t) = x(t)/e_1(t)$; ideally, $y_1(t)$ should have all of its extrema with a unity value. Actually, normalized data still have amplitudes higher than unity occasionally because of the error of the B-spline interpolation;
- Repeat the normalized implementation until the values of normalized data are all less or equal to unity:

$$y_2(t) = y_1(t)/e_2(t), \cdots, y_n(t) = y_{n-1}(t)/e_n(t) \qquad (3.17)$$

- After n times iterations, the normalized data $F(t) = y_n(t)$ and is designated as the empirical FM parts of original IMF. It is a pure frequency modulated function with unity amplitude;
- The AM part is defined as $A(t) = x(t)/y_n(t)$.

It should be noted that the selection of repetition time is a challenging task. If n is big, it influences the computation time; if it is too small, it impacts on the calculation of phase angle.

3.3.2 Calculation of the Instantaneous Frequency

The FM part decomposed from IMF is employed to calculate the instantaneous frequency. Supposing that the FM part is a cosine function $F(t) = \cos \varphi(t)$, the HT form of $F(t)$ can be expressed as follows:

$$F_H(t) = \sin \phi(t) \tag{3.18}$$

Combing the AM part of IMF, the analytical function of the IMF can be obtained.

$$z(t) = A(t)F(t) + jA(t)F_H(t) = A(t)e^{j\varphi(t)} \tag{3.19}$$

Next, the quadrature of $F(t)$ is $\sin \phi(t) = \sqrt{1 - F^2(t)}$. Based on the theory of four-quadrant convention, the signs of FM and its quadrature should be assumed to be the same in the process of quadrature computation. Compared with NHT, normalized direct quadrature (NDQ) has the following advantages:

- NDQ bypasses the limitations of integral interval;
- The calculated value at specified point is not influenced by the neighing points;
- The calculated phase is more exactly than HT.

According to the analysis in [17], there are two methods that can be used to calculate the phase angle after getting the quadrature: one is the **arccosine** function and the other is the **arctangent** function.

$$\begin{cases} \phi(t) = \arccos(F(t)) \\ \phi(t) = \arctan\left(F(t) \Big/ \sqrt{1 - F^2(t)}\right) \end{cases} \tag{3.20}$$

Here, it is noted that there is a necessary condition for the calculation of quadrature. The values in $F(t)$ must be smaller than unity strictly in order to avoid the imaginary part in $\sqrt{1 - F^2(t)}$.

Actually, the calculated values of **arccosine** and **arctangent** are slightly different although they have the same mathematical foundation. In both approaches, unwrap function is added to change absolute jumps greater than or equal to 0 to their π for **arccosine** or $-\pi/2$ to $\pi/2$ for **arctangent** complement. However, both methods still have inherent shortcomings at the extrema and their close neighborhoods. In [17], it introduced a prescribed critical value (default value equals to 0.99). The instantaneous frequencies in the range of preset value are determined by B-spline fitting with the known instantaneous frequencies. The influences of the extrema and their close neighbors are eliminated by the interpolating approximation. Although both **arccosine** and **arctangent** methods are almost equally effective, the application of **arctangent** function is more frequent. Therefore, the **arctangent** is utilized to calculate the phase angle simply. Then the instantaneous frequency of normalized IMF is calculated by the unwrapped phase angle.

3.3.3   Calculation of the Instantaneous Amplitude and Damping Ratio

According to the results of normalized IMF, AM part is expressed as [18]

$$A(t) = x(t)/y_n(t) = e_1(t)e_2(t)\cdots e_n(t) \tag{3.21}$$

where $e_i(t)$ is the envelope of the extrema in the i-th repetition. It means that the instantaneous amplitude is equal to the multiplication of all the envelope signals in the normalized process. Assuming the instantaneous amplitudes at t_i and t_{i+1} are A_i and A_{i+1}, respectively, there is the following relationship between $Ae^{\sigma t_{i+1}}$ and $Ae^{\sigma t_i}$:

$$Ae^{\sigma t_{i+1}} = Ae^{\sigma(t_i + \Delta t)} = Ae^{\sigma t_i}e^{\sigma \Delta t} \tag{3.22}$$

where A is the initial amplitude and σ is the damping ratio. Then

$$Ae^{\sigma t_{i+1}}/Ae^{\sigma t_i} = e^{\sigma \Delta t} \tag{3.23}$$

In this assumption, an approximate estimation of system damping can be obtained simply by taking the natural logarithm of (3.23):

$$\sigma = \ln(A_{i+1}/A_i)/\Delta t \tag{3.24}$$

From (3.24), it is clear that the damping depends on the instantaneous amplitude at t_i and t_{i+1}. Then the damping ratio can be calculated as follows:

$$\xi = -\sigma \left/ \sqrt{(2\pi f)^2 + \sigma^2} \right. \tag{3.25}$$

From (3.22) to (3.25), it provides a choice to calculate the instantaneous damping ratio based on the AM parts and FM parts of the normalized IMF. Therefore, both NHT and NDQ can be used to calculate the oscillatory parameters of the single nonlinear measured signal.

3.4   The Flowchart of the Improved HHT

In order to analyze the dynamic characteristics and to estimate oscillatory parameters of measured signal from single measurement location, an improved HHT is proposed. First, aiming at overcoming the shortcomings of conventional EMD when it is utilized to analyze LFO signal, the improved EMD is presented based on the ESE and FHT. The measured signal is decomposed into a set of IMFs by the improved EMD. Then, IMFs are divided into three groups based on their time–frequency features. In time domain, the energy and mean values of each IMF are considered as

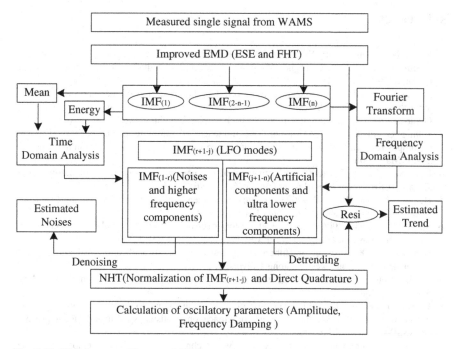

Fig. 3.11 The diagram of improved HHT

the criteria. In frequency domain, the frequency distribution of each IMF is analyzed using FFT. Next, the instantaneous parameters of IMFs, which stand for the interested oscillation mode, are calculated using NHT. The diagram of improved HHT is shown in Fig. 3.11.

3.5   Summary

In this chapter, the shortcomings of traditional HHT when is applied to analyze the LFO problem are analyzed and their corresponding improvements are proposed. Aiming at EEs and mode-mixing phenomena, ESE and FHT are added into the sifting process of classical EMD. In order to overcome the Gibbs problems in HT, NHT is presented to calculate the oscillatory parameters of the IMF which stands for the specific oscillation mode. The improved HHT, which integrates the improved EMD and NHT, is proposed to calculate the oscillatory parameters of the single measured signal. The extension of improved HHT to the synchronized multi-signals from wide area provides the condition for identifying the dominant oscillation mode, node contribution factor (NCF), and approximate mode shape (AMS).

References

1. Huang NE, Shen Z, Long SR et al (1998) The Empirical mode decomposition and the hilbert spectrum for nonlinear and non-stationary time series analysis. Proc R Soc Lond A (454):903–995
2. Rilling G, Flandrin P, Gonçalvès P (2003) On empirical mode decomposition and its algorithms. In: IEEE-EURASIP workshop on nonlinear signal and image process NSIP-03, Grado (I) 2003. http://perso.ens-lyon.fr/patrick.flandrin/emd.html. Accessed 25 Aug 2015
3. Rilling G, Flandrin P (2006) On The Influence of sampling on the empirical mode decomposition. In: IEEE international conference on acoustics, speech and signal process, 14–16 May 2006
4. Peng SL, Hwang WL (2008) Adaptive signal decomposition based on local narrow band signals. IEEE Trans Signal Process 56(7):2669–2676
5. Huang DJ, Zhao JP, Su JL (2003) Practical implementation of the Hilbert–Huang transform algorithm. Acta Oceanol Sinica 25(1):1–11
6. Shu ZP, Yang ZC (2006) A better method for effectively suppressing end effect of empirical mode decomposition (EMD). J Northwest Polytech Univ 24(5):639–643
7. Huang CT (2006) Research on Hilbert–Huang transform and its application. A dissertation for master degree in Southwest Jiaotong University, 2006
8. Lin L, Zhou T, Yu L (2009) Edge effect process technique in EMD algorithm. Comput Eng 35(23):265–268
9. Deering R, Kaiser JF (2005) The use of a masking signal to improve empirical mode decomposition. In: IEEE international conference on acoustic, speech and signal process (ICASSP05), 18–23 Mar 2005
10. Senroy N, Suryanarayanan S (2007) Two techniques to enhance empirical mode decomposition for power quality applications. In: IEEE power engineering society general meeting, 24–28 June 2007
11. Fledman M (2009) Analytical basics of the EMD: two harmonics decomposition. Mech Syst Signal Process 23(7):2059–2071
12. Laila DS, Messina AR, Pal B (2009) A refined Hilbert–Huang transform with applications to inter-area oscillation monitoring. IEEE Trans Power Syst 24(2):610–619
13. Mu G, Shi KP, An J et al (2008) Signal energy method based on EMD and its application to research of low frequency oscillation. Proc CSEE 28(19):36–41
14. Poon KP, Lee KC (1988) Analysis of transient stability swings in large interconnected power systems by Fourier transformation. IEEE Trans Power Syst 3(4):1573–1581
15. Messina AR, Vittal V, Heydt GT et al (2009) Non-stationary approaches to trend identification and denoising of measured power system oscillation. IEEE Trans Power syst 24(4):1798–1807
16. Flandrin P, Rilling G, Gonçalvés P (2004) Empirical mode decomposition as a filter bank. IEEE Signal Process Lett 11(2):112–114
17. Huang NE, Wu ZH, Arnold KC et al (2009) On instantaneous frequency. Adv Adapt Data Anal 1(2):177–229
18. Yang DC, Rehtanz C, Li Y et al (2012) A hybrid method and its applications to research of low frequency oscillations. IEEE/PES Transm Distrib Conf Exposition 2012:7–10
19. Haykin S (2000) Communication system, 4th edn. Wiley, NJ. ISBN:0-471-17869-1

Chapter 4
Oscillation Model Identification Based on Nonlinear Hybrid Method (NHM)

The oscillatory parameters of each synchronized measured signal can be calculated respectively by improved HHT according to the Chap. 3. However, the improved HHT cannot provide the spatial relationships among all the measured signals. In this chapter, the relative phase calculation algorithm (RPCA) is presented to explore the spatial distribution of the specific oscillation mode. Moreover, the concepts of NCF and AMS are proposed to describe the phase information of specific oscillation mode based on the multi-measured signals.

4.1 Identification of Dominant Oscillation Mode

It has been shown that there are $N - 1$ oscillation modes when the number of generators is N. This criterion can be extended to the number of measured signals. In fact, although there are many oscillation modes in actual power system, the dynamic responses are usually controlled by a few of modes, especially the dominant oscillation mode. Furthermore, the triggered oscillation modes are limited under a specific disturbance. Therefore, the problem of selecting the most significant mode has been studied intensively by many researchers. In the conventional model-based method, dominant oscillation mode can be determined by eigenvalues and eigenvector. In this chapter, the dominant oscillation mode is defined based on the calculated oscillatory parameters using the improved HHT. The process of identification of specific oscillation mode is summarized as follows:

1. Decompose the single measured signal from WAMS into corresponding IMFs by the improved HHT, respectively;
2. For the dominant oscillation mode, the IMF with the biggest mean value of oscillation amplitude is selected as the reference. Here, are two meanings: one is that the initial amplitude of selected IMF is big and the other is that the damping of selected is near to zero or negative;

3. Locate the node i which contains the selected IMF;
4. Calculate the mean frequency of the selected IMF and is set as $f_{i,1}$;
5. Define the threshold $f_{\varepsilon 1}$ according to the value of $f_{i,1}$. Usually, $f_{\varepsilon 1}$ is set as $10\% \times f_{i,1}$;
6. Compare the frequencies of other IMFs, if the frequency of the IMF satisfies the following condition: $f_{i,1} - f_{\varepsilon 1} \le f \le f_{i,1} + f_{\varepsilon 1}$, this node which contains this IMF is considered as involving in the dominant oscillation mode;
7. Repeat the comparisons in Step 6 for all the nodes;
8. Identify the nodes which participate in the dominant oscillation mode;
9. Subtract IMFs which belong to the dominant mode;
10. For the left IMFs, repeat Steps 2–9 and extract IMFs which participate in the other oscillation modes.

The process to identify the dominant oscillation mode based on the multi-measured signals and improved HHT is displayed in Fig. 4.1.

For the specific node, the number of oscillation mode is finite. Thus, IMFs which belong to the specific mode are finite. It should be noticed that only one IMF can be considered as in the specific oscillation mode for each measured signal. The number of measured signals which are involved in the specific oscillation mode is less than the number of nodes. The grouping results of IMFs are shown in Fig. 4.2.

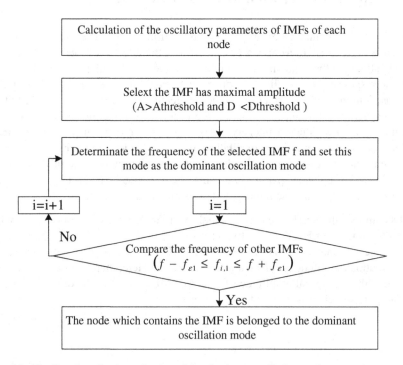

Fig. 4.1 The flowchart for determination of the dominant oscillation mode

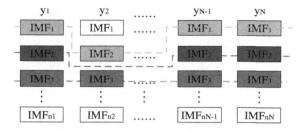

Fig. 4.2 The identification results of different oscillation modes

4.2 The Processing of Oscillation Mode Identification

4.2.1 Calculation of the Absolute Phase (AP) and Relative Phase (RP) of IMF

The oscillatory parameters of each IMF which are involved in the specific oscillation mode can be calculated by improved HHT. In this section, the phase relationships among IMFs are studied. According to the definition, two characteristics of IMF are summarized as follows: (1) the number of extreme in one IMF is defined; (2) IMF is symmetrical with X-axis. These two characteristics provide a method to explore the phase information of IMFs. Figure 4.3 shows a typical IMF.

Apparently, the phase difference of IMFs plays an important role in the node grouping. In [1], the angle information is used to calculate the absolute angle of data point for the specific IMF. However, there are some sudden burrs in the extrema and their neighboring points during calculation of the relative angle. Here, the phase information is utilized to improve the calculation accuracy. The steps are described as follows:

1. The absolute phases (APs) of IMF are determined by hypothesis method. For the specific IMF, the phase angle of positive zero-crossing is set as 0 (or 2π) and the negative zero-crossing is defined as π; the maximum points and minimum points are $\pi/2$ and $3\pi/2$, respectively. The APs of the data points between

Fig. 4.3 Typical IMF with its extrema and zero-crossings

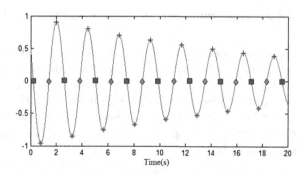

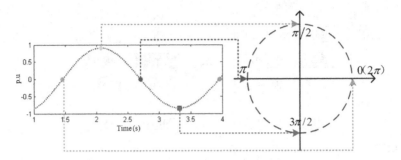

Fig. 4.4 The sketch of calculation absolute phase

zero-crossings and the extrema can be obtained by equipartition of $\pi/2$. The sketch of calculation AP is shown in Fig. 4.4. In this plot, the green dot stands for the positive zero-crossings while the blue dot means the negative zero-crossing. The red square is the minimal points and the cyan diamond indicates the maximal point.

2. The calculated Aps of IMFs are unwrapped by 2π, respectively.
3. For two IMFs, the relative phase (RP) is defined as the point phase difference between unwrapped APs. For example, if the lengths of two IMFs are both M, the unwrapped AP for the ith data point of IMF1 and IMF2 is θ_{1i} and θ_{2i}, respectively, then the RP between IMF1 and IMF2 is calculated by the following formula.

$$\Phi_i^{12} = \sum_{i=1}^{M} \frac{\theta_{1i} - \theta_{2i}}{M} \tag{4.1}$$

AP and RP calculation provide phase relationships among IMFs which are involved in the specific oscillation mode. The AMS of the dominant oscillation mode can be determined by the oscillatory amplitudes and RPs [1].

4.2.2 Determination of Node Contribution Factor (NCF)

Based on the calculated oscillatory parameters of IMFs, the node which participates in the same oscillation mode can be identified. Furthermore, the calculation of the RP provides the phase relationships among IMFs. In this section, the concept of node contribution factor (NCF) is proposed by combining the oscillatory parameters and phase relationships. Obviously, for each node, NCF expresses its contribution degree to the specific oscillation mode. The steps to determine NCFs are explained as follows [2]:

1. For the dominant oscillation mode λ_1, the IMF with the biggest mean amplitude is set as the reference and its AP is set as 0. Then, the RPs between reference and other IMFs are calculated by (4.1) and expressed as $\Phi^{\lambda_1-k}(k \in 1, 2, \ldots, N_1 - 1)$. N_1 is the number of IMFs which are involved in the dominant oscillation mode and k is the corresponding order;
2. The oscillation energy of the IMF with the biggest mean amplitude is calculated as

$$P^{\lambda_1-\max} = A^{\lambda_1-\max} \cos(0) \tag{4.2}$$

   A is the mean value of oscillatory amplitude in specific time range.
3. The oscillation energy of other IMFs is determined in the following forms:

$$P^{\lambda_1-k} = A^{\lambda_1-k} \cos(\Phi^{\lambda_1-k}) \tag{4.3}$$

   A^{λ_1-k} is the mean value of oscillatory amplitude of kth IMF in the same time range as mentioned above.
4. The NCF of kth IMF is calculated based on the maximal oscillation energy and its oscillation energy.

$$C^{\lambda_1-k} = \frac{P^{\lambda_1-k}}{P^{\lambda_1-\max}} = \frac{A^{\lambda_1-k} \cos(\Phi^{\lambda_1-k})}{A^{\lambda_1-\max} \cos(0)} (k \in 1, 2, \ldots, N_1) \tag{4.4}$$

5. Similarly, NCFs of nodes which include the same oscillation mode can be determined by Steps 1–4.

As for the specific mode, the NCF of each node indicates the contribution to this mode. According to the definition, the biggest value of NCF is 1 [2]. However, the definition of NCF is based on the mean values of amplitude and phase. In order to trace the instantaneous variation of the oscillation, the reasonable choice of length of time window is necessary.

4.2.3 Computation of Approximate Mode Shape (AMS)

For the given oscillation mode, its properties are described by its oscillatory parameters and mode shape. Many papers have offered approaches for computing the modal parameters from the measured data. However, only a few publications have considered the mode shape. In fact, the mode shape may provide critical information for control decisions. For example, mode shape can be utilized to optimize the generator and/or load tripping schemes to improve the stability of power system.

According to the theory in [3], the mode shape of LFO is defined as the magnitude and angle of the mode relative to a reference location in the power system. It describes the relative participation of the state variables in a specific mode. Therefore, the concept of the AMS is proposed in this section based on the results of NCFs.

NCF is used to evaluate the contribution of each node to the dominant mode. As for the mode, AMS is defined as all the NCFs of nodes which are involved in this mode. Compared with the standard definition of mode shape, the absolute values of NCFs are similar to the magnitude and the values of RPs are approximate to the relative phases to the reference which is selected based on the maximal oscillatory energy.

4.2.4   Coherency of the Measured Signals

In the multi-machine system, it has been shown that some generators (nodes) tend to swing together after the small or sudden disturbance. The coherency of the nodes can help to identify the oscillation boundaries and to avoid unnecessary tripping. For instance, in [3], it proposed the controlled islanding schemes based on the generator coherency. The stabilities of the islands are guaranteed by the fact that all the generators in these islands are coherent. In this section, in order to locate the oscillation interface, the generators coherence is extended to the nodes coherence.

WAMS provides the opportunity to implement the nodes coherence from two aspects: (1) most of the key nodes including the substations and generators have built up the PMUs which are employed to monitor the dynamic parameters timely; (2) the advanced signal transmission technology ensure that the local and wide area signal are synchronous.

As for the specific oscillation mode, the AP of the reference is set as 0 in the whole data range. It is obvious that the RP between reference and considering node is in the range of $0-2\pi$. According to the values of RP, all the nodes can be divided into two groups. The RP in the range of $\pi/2-3\pi/2$ is considered in the negative group and the RP in the range of $(0-\pi/2) \cup (3\pi/2-2\pi)$ is considered in the positive group [2, 4]. Figure 4.5 shows a conceptual representation of nodes coherence at the same oscillation mode.

Fig. 4.5 The conceptual representation of the nodes coherence

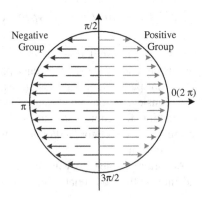

In Fig. 4.5, all the nodes are compared with the reference node. Furthermore, the transmission line which links the nodes in different groups is considered as the key transmission lines. The secant line of all transmission lines is defined as the oscillation interface.

4.2.5   Flowchart of the Nonlinear Hybrid Method (NHM)

As highlighted in the previous sections, improved HHT is employed to calculate the oscillatory parameters for the single measured signal. When data is available at multiple locations, the multi-measured signals can be obtained timely and synchronously. It is necessary to set up a model that represents system behavior.

Supposing the measured signal is available at N location and the length of time window is M, the measured data in y_j $(j = 1, 2, ..., N)$ at t_k $(k = 1, 2, ..., M)$ is expressed as $a(y_j, t_k)$. Obviously, y_j denotes the spatial variables and t_k indicates the temporal variables. Using this notation, the set of measured data is expressed as a matrix $\mathbf{Y}(\mathbf{y}, \mathbf{t})$.

$$
\begin{aligned}
\mathbf{Y}(\mathbf{y}, \mathbf{t}) &= \begin{bmatrix} \mathbf{a}(\mathbf{y}_1, \mathbf{t}) & \mathbf{a}(\mathbf{y}_2, \mathbf{t}) & \cdots & \mathbf{a}(\mathbf{y}_N, \mathbf{t}) \end{bmatrix} \\
&= \begin{bmatrix}
a(y_1, t_1) & a(y_2, t_1) & \cdots & a(y_N, t_1) \\
a(y_1, t_2) & a(y_2, t_2) & \cdots & a(y_N, t_2) \\
\vdots & \vdots & \ddots & \vdots \\
a(y_1, t_M) & a(y_2, t_M) & \cdots & a(y_N, t_M)
\end{bmatrix}
\end{aligned}
\tag{4.5}
$$

The size of $\mathbf{Y}(\mathbf{y}, \mathbf{t})$ is $M \times N$. In this matrix, the column means the system response to all the events in the specific time window; the row represents the system response to a single event at all the locations. The single spatial variable $\mathbf{a}(\mathbf{y}_j, \mathbf{t})$ is considered as one measured signal and its oscillatory parameters are calculated by the improved HHT. The former part of this chapter has proposed a series of algorithms for exploring the relationships among the spatial variables.

$\mathbf{Y}(\mathbf{y}, \mathbf{t})$ is transformed into the following form by applying improved EMD to each spatial variable.

$$
\begin{aligned}
\mathbf{Y}(\mathbf{y}, \mathbf{t}) &= \begin{bmatrix} \mathbf{IMF}(\mathbf{y}_1, \mathbf{t}) & \mathbf{IMF}(\mathbf{y}_2, \mathbf{t}) & \cdots & \mathbf{IMF}(\mathbf{y}_N, \mathbf{t}) \end{bmatrix} \\
&= \begin{bmatrix}
IMF_{11} & IMF_{21} & \cdots & IMF_{N1} \\
IMF_{12} & IMF_{22} & \cdots & IMF_{N2} \\
\vdots & \vdots & \ddots & \vdots \\
IMF_{1n_1} & IMF_{2n_2} & \cdots & IMF_{Nn_N}
\end{bmatrix}
\end{aligned}
\tag{4.6}
$$

Each IMF is the function of time, IMF_{jn_j} stands for n_jth IMF in jth spatial variable.

For each time series, the number of IMFs is finite. According to the proposed method in Sect. 3.2, IMFs which means the physical oscillation mode is extracted based on the signal energy analysis and frequency distribution. The detailed analysis methods have been introduced in Chap. 3. Next, the extracted IMFs in each time series are researched from two aspects: one is the calculation of the instantaneous oscillatory parameters based on improved HHT; the other is determination of the RPs based on RPCA. The calculated amplitudes and frequencies of IMFs are used to identify the specific oscillation mode. The IMFs which are involved in the dominant oscillation mode are estimated by the calculated frequencies and the threshold. The matrix of the IMFs in dominant oscillation mode can be expressed as follows:

$$
\mathbf{IMFs_1} = \begin{bmatrix} \vdots \\ \text{IMF}_{i,j} \\ \vdots \end{bmatrix} (i \in [1,2,\ldots,M], j \in ([1,\ldots,n_1],[1,\ldots,n_2],\ldots,[1,\ldots,n_M]))
$$

(4.7)

where $\text{IMF}_{i,j}$ is the jth IMF in ith measured signal.

For the dominant oscillation mode, the AP of $\text{IMF}_{i,j}$ with biggest energy is set as the reference and the RPs of other IMFs are determined by

$$
\mathbf{Y} = [\mathbf{y_1}\ldots\mathbf{y_N}] = \begin{bmatrix} y_1(t_1) & y_2(t_1) & \cdots & y_N(t_1) \\ \vdots & \vdots & \ddots & \vdots \\ y_1(t_i) & y_2(t_i) & \cdots & y_N(t_i) \\ \vdots & \vdots & \ddots & \vdots \\ y_1(t_M) & y_2(t_M) & \cdots & y_N(t_M) \end{bmatrix}
$$

(4.8)

Then, the NCF is calculated by the amplitudes and RPs. The AMS of dominant oscillation mode is formed based on all the NCFs. According to the grouping results, the key transmission lines and oscillation interface can be identified. Generally speaking, most nodes of interconnected power system have been installed with PMUs, especially in the output of power plants and the main substations. Thus, the dynamic responses of most nodes can be recorded and transferred to the phasor data concentrator (PDC) timely and synchronously. Under this background, the flowchart of the proposed NHM based on the multi-measured signals is shown in Fig. 4.6.

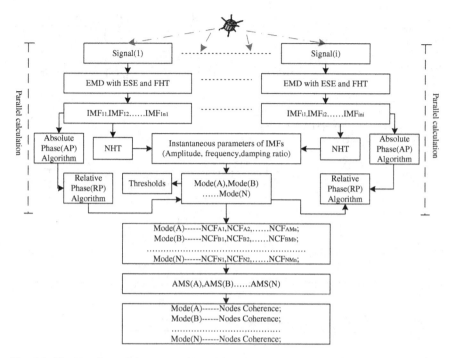

Fig. 4.6 The flowchart of the proposed NHM based on multi-measured signals

4.3 Study Case

In order to demonstrate the ability of proposed NHM to analyze the simulation data, EPRI-36 system is built in the Power System Analysis Software Package (PSASP). Actually, this system has been widely used in many publications as a research demonstration model for LFO analysis. This model includes 8 generators and 36 buses. The single diagram is shown in Fig. 4.7.

The small disturbance and the calculation conditions are set as follows: the length of time window is 10 s; the sampling frequency is 100 Hz; an impact load is added on the bus 19 at 0.12 s. The oscillatory parameters and mode shapes of the seven oscillation modes are calculated by the small signal analysis blocks, the calculated parameters and corresponding mode shapes are shown in Fig. 4.8 [5] and Table 4.1, respectively.

Under this disturbance, the output active power of each generator (the reference active power is 100 MW) and their FFT spectrums are shown in Fig. 4.9.

The log-scale plot is employed in order to display the results more clearly. In Fig. 4.10b, there are two peaks in the FFT spectrum: one is 0.7813 Hz, the other is 1.6602 Hz. Then, the improved EMD is utilized to decompose the measured signals. The IMF1 and IMF2 results are shown in Fig. 4.10.

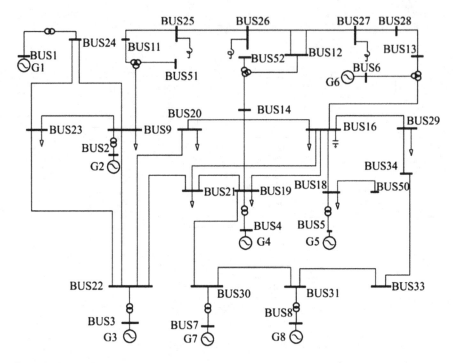

Fig. 4.7 EPRI—36 bus pure AC power system

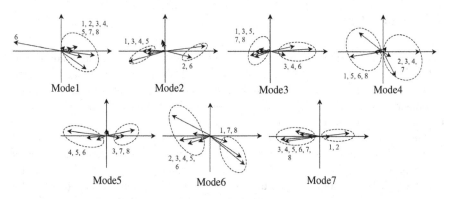

Fig. 4.8 The mode shapes of 7 oscillation modes

It is obvious that the existing time of IMF1 is 0–5 s while IMF2 is 1–10 s. Furthermore, the frequency of IMF1 is higher than IMF2. According to the theory of EMD and the analysis in Chap. 3, it can produce the mode-mixing phenomena in the sifting process. However, the improved EMD can reduce the influence of mode-mixing and EEs effectively. That can be seen from the decomposed results.

Table 4.1 The results of eigenvalues computation

Mode	Real part (s^{-1})	Image part (rad/s)	Frequency (Hz)	Damping ratios (%)
1	−6.7742	14.6082	2.3249	42.0694
2	−0.7904	11.4707	1.8256	6.8747
3	−0.9025	10.3686	1.6502	8.6713
4	−0.6177	7.8593	1.2508	7.8356
5	−0.6868	7.3088	1.1632	9.3551
6	−0.3002	6.2393	0.9931	4.8057
7	−0.0486	4.9263	0.7841	0.9861

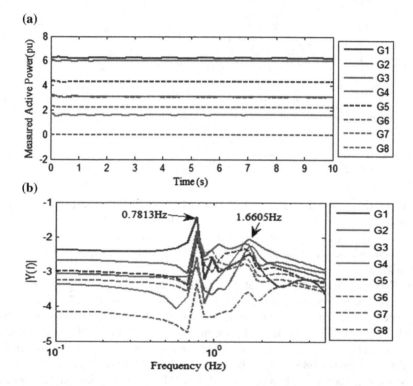

Fig. 4.9 The measured active power and FFT spectrum of each generator. **a** The output active power of each generator. **b** The FFT spectrum of output active power

Next, the signal energies and frequency distributions of IMF1 and IMF2 are displayed in Fig. 4.11, respectively.

From Fig. 4.11, it can be seen that the frequency of IMF1 of each generator is nearly to 1.66 Hz and the frequency of IMF2 is about 0.78 Hz. Both IMF1 and IMF2 can be considered as the mono-frequency oscillation component.

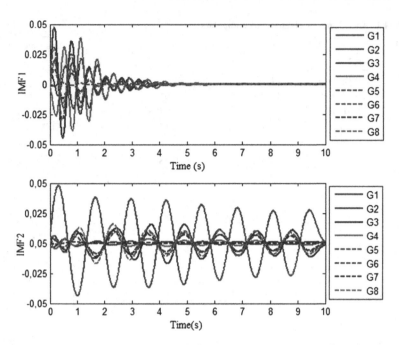

Fig. 4.10 The decomposed results (IMF1 and IMF2)

Furthermore, as for Mode 1 ($f_2 = 0.78$ Hz), IMF2 of G1 has biggest signal energy; as for Mode 2 ($f_1 = 1.66$ Hz), IMF1 of G4 has biggest signal energy.

Next, the NHT is employed to calculate the oscillatory parameters and the RPCA is utilized to determine the phase relations of the IMFs. The results are listed in Table 4.2.

In Table 4.2, it can be seen that: (1) the IMF2 of G1 has the biggest oscillation amplitude and its frequency is 0.77 Hz. If the threshold is set as 0.77×5 % Hz, all the generators can be considered to involve in the Mode 1; (2) besides the IMFs which involved in the Mode 1, the mean amplitude of IMF1 of G4 is biggest. Therefore, G4 is selected as the reference for the Mode 2. The threshold of the oscillation mode is 1.686×10 % Hz. It is clear that all the generators participate in the Mode 2; (3) according to RPCA, the RPs of IMFs are determined and the results provide criteria for the node grouping; (4) from the damping ratios, all the damping ratios of IMF2 are in the range of 0.5–3 %. Therefore, the Mode 1 can be considered as the dominant oscillation mode. Moreover, Mode 2 is an obvious decreased amplitude oscillation mode.

The NCFs of each generator in Mode 1 and Mode 2 are calculated based on the oscillatory parameters and RPs. The calculated results are displayed in Fig. 4.12.

In order to evaluate the accuracy of proposed method, the NCFs of every generator in the mode shapes are compared with the order of magnitude (dB) values in Fig. 4.9b. The comparison results are listed in Table 4.3.

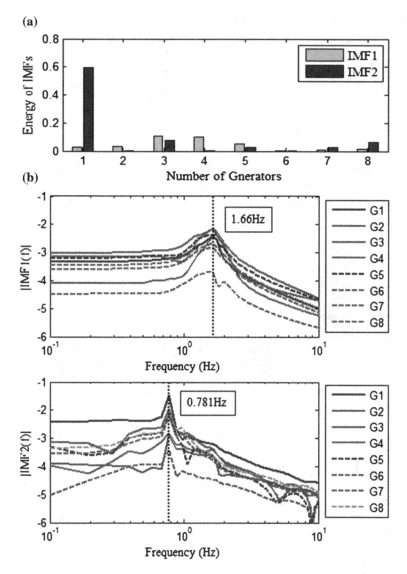

Fig. 4.11 The time–frequency analysis of IMFs. **a** The signal energy of IMF. **b** The frequency distribution of IMF

From Table 4.3, the NCFs of each generator are the same as the order of magnitude values in Fig. 4.9b. Furthermore, based on the NCFs and RPs, the AMS of Mode 1 (dominant oscillation modes) is shown in Fig. 4.13a. Similarly, the AMS of Mode 2 (1–5 s) is displayed in Fig. 4.13b.

Table 4.2 The oscillatory parameters of IMFs and relative phase relationships

Name	Amplitude (p.u)		Frequency (Hz)		Damping ratios (%)		RP (θ)	
	IMF2	IMF1	IMF2	IMF1	IMF2	IMF1	IMF2	IMF1
G1	0.0317	0.0064	0.779	1.625	1.224	6.37	0	3.286
G2	0.0012	0.0086	0.787	1.670	1.547	5.97	6.230	0.485
G3	0.0128	0.0091	0.784	1.646	1.211	7.84	2.873	3.338
G4	0.0023	0.0131	0.774	1.682	0.763	5.84	3.109	0
G5	0.0078	0.0067	0.780	1.645	0.936	8.28	2.956	3.222
G6	0.0004	0.0003	0.773	1.670	0.919	7.05	3.023	−0.302
G7	0.0069	0.0021	0.784	1.648	0.751	8.61	3.171	3.193
G8	0.0104	0.0029	0.784	1.647	0.652	8.65	3.159	3.258

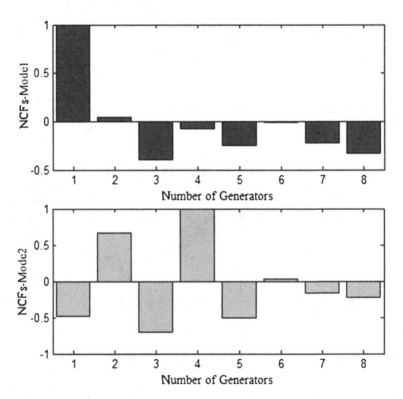

Fig. 4.12 The NCF of each generator in Mode 1 and Mode 2

As for the Mode 1, it is expressed as the positive group (G1, G2) oscillating against negative group (G3, G4, G5, G6, G7, G8). As for the Mode 2, it also involves all the generators, but it has different grouping results and it is expressed as the positive group (G2, G4, G6) and negative group (G1, G3, G5, G7, G8).

Table 4.3 The comparison of NCFs and magnitudes of FFT spectrum

Name	Results of FFT		Results of AMS	
	Mode 1	Mode 2	Mode 1	Mode 2
G1	1	5	1.0000	−0.4944
G2	7	3	0.0373	0.6373
G3	2	2	−0.4088	−0.8476
G4	6	1	−0.0712	1.0000
G5	4	4	−0.2464	−0.6049
G6	8	8	−0.0117	0.0306
G7	5	7	−0.2172	−0.2086
G8	3	6	−0.3275	−0.2826

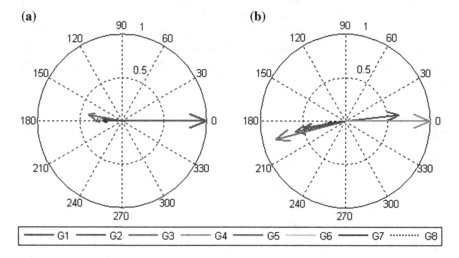

Fig. 4.13 The AMSs of Mode 1 (a) 0.78 Hz and Mode 2 (b) 1.66 Hz

The results identified by NHM are compared with the eigenvalues analysis in Fig. 4.8 and Table 4.1. The calculated Mode 1 is consistent with the Mode 7 and Mode 2 is similar to the Mode 3 which is determined by the model analysis. It is clear that both of the methods have the same grouping results. Although there are a little difference in phase angles and NCFs, the calculated results are convincible. Actually, there are some differences between the theoretical values and real response to the random disturbance in the actual power system. Therefore, it is necessary to monitor and to control the oscillation mode real time.

4.4 Summary

In this chapter, the RPCA is presented to explore the phase relations among IMFs involved in the same oscillation mode. The concepts of NCF and AMS are proposed to describe the phase information of specific oscillation mode based on the multi-measured signals. According to the descriptions in Chaps. 3 and 4, a novel method, named as NHM, is given by combing the improved HHT and RPCA. The NHM can not only provide the oscillatory parameters of single measured signal and NCFs of every mode, but also calculate the AMSs of the oscillation modes. It links the gap between the oscillatory parameters of single measured signal and oscillation mode shape in the ensemble measured matrix.

NHM provides a new path to study the oscillation mode from the nonlinear viewpoint based on the measured signals in wide area. The eigenvalues analysis has shown that the calculated results of proposed method are feasible and credible. However, the evaluation method does not break the limitation of the system model. Therefore, it is necessary to build up an evaluation method based on the measured signals.

References

1. Duan G, Lin JJ, Wu JT (2009) Analysis method of node contribution factor of low frequency oscillation based on wide area measurement information. CN101408577A:4–15
2. Li J, Xue AC, Wang JP et al (2010) A new node contribution factors for the low frequency oscillations of power system based on the PMU's data and HHT. In: 5th international conference on critical infrastructure (CRIS), 20–22 Sept 2010
3. Rogers G (2000) Power system oscillations. Kluwer Academic Pub, Norwell (USA). ISBN 0-7923-7712-5
4. Yang DC, Rehtanz C, Li Y et al (2011) A novel method for analyzing dominant oscillation mode based on improved EMD and signal energy algorithm. Sci China Ser E-Tech Sci 54 (9):2493–2500
5. Cai GW, Yang DY, Zhang JF et al (2011) Mode identification of power system low frequency oscillation based on measured signal. Power Syst Technol 35(1):59–65

Chapter 5
Identification of Dominant Complex Orthogonal Mode (COM)

The essence of LFO can be considered as the energy distribution in time and space. Therefore, characterization of LFO from the aspects of temporal and spatial has great theoretical and practical significance for power system stability analysis. In this chapter, the complex orthogonal decomposition (COD) is introduced and utilized to analyze temporal and spatial characteristics of the oscillation models in power systems based on the time-synchronized ensemble measurement matrix [1]. And the actual measured data from wide-area-protector (WAProtector) is employed to validate the effectiveness of the near real-time application of the Complex Orthogonal Decomposition-Sliding Window Recursive Algorithm (COD-SWRA).

5.1 Introduction of Spatial and Temporal Behaviors of Oscillation Mode

The construction of WAMS provides the opportunity to obtain the time-synchronized data from wide area. At instance of time t_j, the measured information $\mathbf{u}(t_j)$ from location $y_i(i \in (1, 2, \ldots N))$ can be represented as:

$$\mathbf{u}(t_j) = \left[u(y_1, t_j), \ldots, u(y_N, t_j) \right] \tag{5.1}$$

If $N = 10$, the random energy distribution can be indicated as shown in Fig. 5.1.

Assuming that the length of time series is M, the ensemble measurement matrix can be expressed as follows [2]:

$$\mathbf{X} = [\mathbf{u_1} \ldots \mathbf{u_N}] = \begin{bmatrix} u(y_1, t_1) & \cdots & u(y_N, t_1) \\ \vdots & \ddots & \vdots \\ u(y_1, t_M) & \cdots & u(y_N, t_M) \end{bmatrix} \tag{5.2}$$

$u(y_i, t_j)$ represents the value of the field at time t_j and spatial position y_i.

© Springer-Verlag Berlin Heidelberg 2016
Y. Li et al., *Interconnected Power Systems*, Power Systems,
DOI 10.1007/978-3-662-48627-6_5

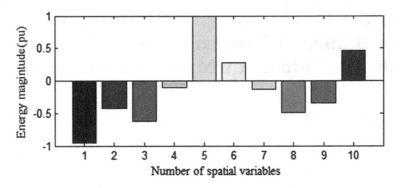

Fig. 5.1 Illustration of the energy spatial distribution

In Eq. (5.2), the response of the system to a single event at the specified time is expressed as one row. Measured information which contains the ensemble of time responses to multiple events at one location is represented as one column. The conceptual representation of spatial and temporal variability of oscillation mode in power system is depicted in Fig. 5.2.

The essence of the COD analysis is to find a spatial and temporal basis that spans an ensemble of data collected from experiments or numerical simulations. The method essentially decomposes a fluctuating field into a weighted linear sum of spatial orthogonal modes and temporal orthogonal modes such that the projection onto the first few modes is optimal. Before introducing the methods of COD, the

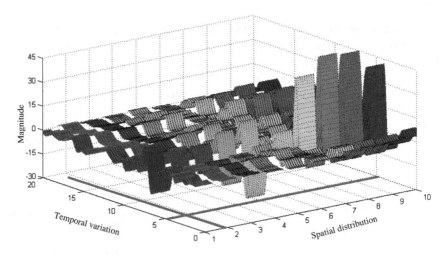

Fig. 5.2 Three-dimensional view of spatial and temporal behavior

trending and the fluctuating part of the ensemble measurement matrix should be separated and the complex ensemble measurement matrix should be built.

5.2 Construction of the Complex Ensemble Measurement Matrix

In order to improve the ability of exploring the temporal dynamic behaviors, every column is separated into the trending part as well as the fluctuating part based on the improved EMD and time-frequency domain analysis of IMFs which are introduced in Chap. 3 [3].

$$\mathbf{u}(y_i, t_j) = \mathbf{u}_m(y_i, t_j) + \widehat{\mathbf{u}}(y_i, t_j) \tag{5.3}$$

where $\mathbf{u}_m(y_i, t_j)$ and $\widehat{\mathbf{u}}(y_i, t_j)$ are the trending part and fluctuating part of ith time series, respectively.

The fluctuating part of original matrix is rewritten as

$$\widehat{\mathbf{Y}} = [\widehat{\mathbf{y}}_1 \ldots \widehat{\mathbf{y}}_N] = \begin{bmatrix} \widehat{u}(y_1, t_1) & \cdots & \widehat{u}(y_N, t_1) \\ \vdots & \ddots & \vdots \\ \widehat{u}(y_1, t_M) & \cdots & \widehat{u}(y_N, t_M) \end{bmatrix} \tag{5.4}$$

According to Eq. (5.4), each column of $\widehat{\mathbf{Y}}$ is set as one unit $\widehat{\mathbf{u}}_j(t) . \widehat{\mathbf{u}}_j(t)$ is the real part of a complex expression $\psi_j(t)$, that is,

$$\widehat{\mathbf{u}}_j(t) = \mathbf{RE}[\psi_j(t)] \tag{5.5}$$

where

$$\psi_j(t) = \widehat{\mathbf{u}}_j(t) + i\xi_j(t) \tag{5.6}$$

Then $\xi_j(t)$ is the HT form of $\widehat{\mathbf{u}}_j(t)$ according to the introduction in Chap. 3:

$$\xi_j(t) = \mathrm{H}[\widehat{\mathbf{u}}_j(t)] = \frac{1}{\pi} \int\limits_{-\infty}^{\infty} \frac{u(l)}{t-l} dl \tag{5.7}$$

HT is equivalent to a phase shift of $\pm\pi/2$ in the spectral domain. Therefore, the complex ensemble measurement matrix can be built as follows:

$$\widehat{\mathbf{Y}}_C = \widehat{\mathbf{Y}} + \mathbf{i} * \widehat{\mathbf{Y}}_H (\mathbf{i} = \sqrt{-1}) \tag{5.8}$$

Then, the complex ensemble measurement matrix can be expressed as in the detailed form of vectors:

$$\hat{\mathbf{Y}}_\mathbf{C} = [\boldsymbol{\psi_1}, \boldsymbol{\psi_2}, \ldots \boldsymbol{\psi_N}] = \begin{bmatrix} \hat{u}(y_1,t_1) + i\xi(y_1,t_1) & \cdots & \hat{u}(y_N,t_1) + i\xi(y_N,t_1) \\ \vdots & \ddots & \vdots \\ \hat{u}(y_1,t_M) + i\xi(y_1,t_M) & \cdots & \hat{u}(y_N,t_M) + i\xi(y_N,t_M) \end{bmatrix}$$

$$(5.9)$$

5.3   Implementations of Complex Orthogonal Decomposition (COD)

Actually, the oscillation processes of the power system in wide area are derived from the nonlinear interactions between different modes of varying spatial scales and temporal frequencies. COD has been applied to decompose the signal into traveling and standing waves and to analyze their spatiotemporal characteristics in many fields, such as the fluids and plasmas. In this section, three different approaches are proposed to implement the COD. The brief introductions on complex eigenvalues decomposition (C-ED) and the complex singular value decomposition(C-SVD) are given firstly. In fact, similar ideas have been also explored by other literatures [1, 4]. However, the mentioned publications do not explain how to further apply the calculated parameters to interpret the LFO phenomenon. Moreover, a novel decomposition method, named augmented matrix decomposition (AMD), is also proposed to realize the COD. In addition, the performances of C-ED, C-SVD, and AMD are compared under four different magnitudes of ensemble measurement matrixes.

5.3.1   Complex Eigenvalues Decomposition (C-ED)

According to Eq. (5.9), the covariance matrix of $\hat{\mathbf{Y}}_\mathbf{C}$ is calculated by the following forms:

$$\begin{cases} \mathbf{F_1} = \hat{\mathbf{Y}}_\mathbf{C}\hat{\mathbf{Y}}_\mathbf{C}^H \big/ M \\ \mathbf{F_2} = \hat{\mathbf{Y}}_\mathbf{C}^H\hat{\mathbf{Y}}_\mathbf{C} \big/ N \end{cases}$$

$$(5.10)$$

The superscript H denotes the conjugate transpose. Evidently, both $\mathbf{F_1}$ and $\mathbf{F_2}$ are Hermitain matrixes and they can be expressed by the following forms, respectively.

$$\begin{cases} \mathbf{F}_1 = \frac{1}{M}\left[(\hat{\mathbf{Y}}\hat{\mathbf{Y}}^T + \hat{\mathbf{Y}}_H\hat{\mathbf{Y}}_H^T) + \mathbf{i} * (\hat{\mathbf{Y}}_H\hat{\mathbf{Y}}^T + \hat{\mathbf{Y}}\hat{\mathbf{Y}}_H^T)\right] = \frac{1}{M}[\mathbf{A} + \mathbf{B} * \mathbf{i}] = \mathbf{U}_1\mathbf{\Lambda}_1\mathbf{V}_1^H \\ \mathbf{F}_2 = \frac{1}{N}\left[(\hat{\mathbf{Y}}^T\hat{\mathbf{Y}} + \hat{\mathbf{Y}}_H^T\hat{\mathbf{Y}}_H) + \mathbf{i} * (\hat{\mathbf{Y}}^T\hat{\mathbf{Y}}_H + \hat{\mathbf{Y}}_H^T\hat{\mathbf{Y}})\right] = \frac{1}{N}[\mathbf{C} + \mathbf{D} * \mathbf{i}] = \mathbf{U}_2\mathbf{\Lambda}_2\mathbf{V}_2^H \end{cases}$$

$$(5.11)$$

The subscript H means the function of HT. From Eq. (5.11), it is clear that $\mathbf{A}$ and $\mathbf{C}$ are symmetric matrix and their eigenvector are real while $\mathbf{B}$ and $\mathbf{D}$ are asymmetric matrix and their eigenvectors are in the form of complex conjugate. Therefore, both $\mathbf{F}_1$ and $\mathbf{F}_2$ have the complex eigenvectors and real eigenvalues. $\mathbf{U}_1$ and $\mathbf{V}_1$ are the right and left eigenvectors of $\mathbf{F}_1$ and $\mathbf{\Lambda}_1 = \mathbf{diag}(\lambda_1, \lambda_2, \ldots \lambda_M)$, respectively. The eigenvalues are real and can be ordered as $\lambda_1 \geq \lambda_2 \geq \cdots \geq \lambda_M \geq 0$. $\mathbf{U}_2$ and $\mathbf{V}_2$ are the right and left eigenvectors of $\mathbf{F}_2$ and $\mathbf{\Lambda}_2 = \mathbf{diag}(\gamma_1, \gamma_2, \ldots \gamma_N)$, respectively. The eigenvalues are real and can also be ordered in the form of $\gamma_1 \geq \gamma_2 \geq \cdots \gamma_N \geq 0$.

In order to quantify the contribution of each mode to the total field, the energy of a vector sequence building a matrix is defined by the Frobenius norm. Then, the total energy of $\mathbf{F}_1$ and $\mathbf{F}_2$ can be determined by Eq. (5.12).

$$\begin{cases} \mathbf{E}^1 = \sum_{i=1}^{M} \lambda_i \\ \mathbf{E}^2 = \sum_{i=1}^{N} \gamma_i \end{cases}$$

$$(5.12)$$

The associated percentage of total energy contributed by each mode is expressed as

$$\begin{cases} \mathbf{E}_i^1 = \lambda_i / \mathbf{E}^1 = \lambda_i \Big/ \sum_{i=1}^{M} \lambda_i \\ \mathbf{E}_i^2 = \gamma_i / \mathbf{E}^2 = \gamma_i \Big/ \sum_{i=1}^{N} \gamma_i \end{cases}$$

$$(5.13)$$

5.3.2 Complex Singular Value Decomposition (C-SVD)

Using the SVD and the form of analytical function, the complex matrix $\hat{\mathbf{Y}}_C$ can be expressed as follows:

$$\hat{\mathbf{Y}}_C = \mathbf{U}\mathbf{\Sigma}\mathbf{V}^H \qquad (5.14)$$

where, the superscript H indicate the conjugate of $\mathbf{V}$, $\mathbf{V}$ is an orthogonal $N \times N$ matrix whose columns are the right singular vectors of $\hat{\mathbf{Y}}_C$; $\mathbf{U}$ is an orthogonal $M \times M$ matrix whose columns are the left singular vectors of $\hat{\mathbf{Y}}_C$; $\mathbf{\Sigma}$ is $M \times N$

matrix containing the singular values of $\hat{\mathbf{Y}}_C$ along the main diagonal and zeros elsewhere. The correlation matrix $\mathbf{F}_1$ and $\mathbf{F}_2$ can be obtained by

$$\mathbf{F}_1 = \frac{1}{M}\mathbf{U}\boldsymbol{\Sigma}\mathbf{V}^H[\mathbf{U}\boldsymbol{\Sigma}\mathbf{V}^H]^H = \frac{1}{M}\mathbf{U}\boldsymbol{\Sigma}\mathbf{V}\mathbf{V}^H\boldsymbol{\Sigma}^T\mathbf{U}^H = \frac{1}{M}\mathbf{U}\boldsymbol{\Sigma}\boldsymbol{\Sigma}^T\mathbf{U}^H \tag{5.15}$$

$$\mathbf{F}_2 = \frac{1}{N}[\mathbf{U}\boldsymbol{\Sigma}\mathbf{V}^H]^H\mathbf{U}\boldsymbol{\Sigma}\mathbf{V}^H = \frac{1}{N}\mathbf{V}\boldsymbol{\Sigma}^T\mathbf{U}^H\mathbf{U}\boldsymbol{\Sigma}\mathbf{V}^H = \frac{1}{N}\mathbf{V}\boldsymbol{\Sigma}^T\boldsymbol{\Sigma}\mathbf{V}^H \tag{5.16}$$

From Eqs. (5.11) to (5.16), it is clear that the left singular vector of $\hat{\mathbf{Y}}_C$ equals to the eigenvectors of the sample covariance matrix $\mathbf{F}_1$ and the right singular vector of $\hat{\mathbf{Y}}_C$ equals to the eigenvectors of the sample covariance matrix $\mathbf{F}_2$. Furthermore, the eigenvalues of matrix $\mathbf{F}_1$ are the square of the singular values δ_i divided by M and the eigenvalues of matrix $\mathbf{F}_2$ are the square of the singular values δ_i divided by N. The relationships between eigenvalues and singular values are determined as follows:

$$\begin{cases} \lambda_i = \delta_i^2/M \\ \gamma_i = \delta_i^2/N \end{cases} \tag{5.17}$$

5.3.3   Augmented Matrix Decomposition (AMD)

From the above description, it can be seen that both the C-ED and C-SVD can be used to extract the optimal basis which stands for the low-dimensional approximate descriptions of high-dimensional processes. However, the mentioned methods need to calculate two forms of autocorrelation matrix of the complex ensemble measurement matrix. In this subsection, the augmented matrix decomposition (AMD) is proposed to determine the complex singular vectors and singular values of $\hat{\mathbf{Y}}_C$.

The augmented matrix is described as:

$$\mathbf{R} = \begin{bmatrix} \hat{\mathbf{Y}} & \hat{\mathbf{Y}}_H \\ -\hat{\mathbf{Y}}_H & \hat{\mathbf{Y}} \end{bmatrix} \tag{5.18}$$

The SVD of $\mathbf{R}$ can be expressed as follows:

$$\mathbf{R} = \mathbf{U}_3 \sum_3 \mathbf{V}_3^T \tag{5.19}$$

$\mathbf{U}_3$ is an orthogonal $2M \times 2M$ matrix; $\mathbf{V}_3$ is an orthogonal $2N \times 2N$ matrix; $\boldsymbol{\Sigma}_3$ is $2M \times 2N$ matrix and the elements in the diagonal are the singular values and the values in elsewhere are zero.

The SVD in form of $\mathbf{R}$ is rewritten by the form of block matrix.

$$\mathbf{R} = \begin{bmatrix} \hat{\mathbf{Y}} & \hat{\mathbf{Y}}_H \\ -\hat{\mathbf{Y}}_H & \hat{\mathbf{Y}} \end{bmatrix} = \begin{bmatrix} \mathbf{U}_R & \mathbf{U}_I \\ -\mathbf{U}_I & \mathbf{U}_R \end{bmatrix} [\mathbf{W}] \begin{bmatrix} \mathbf{V}_R^H & \mathbf{V}_I^H \\ -\mathbf{V}_I^H & \mathbf{V}_R^H \end{bmatrix} \quad (5.20)$$

The relationships between $\mathbf{U}$ and $\mathbf{U}_R$ as well as $\mathbf{U}_I$ can be deduced based on Eqs. (5.19) and (5.20). Similarly, the relationships between $\mathbf{V}$ and $\mathbf{V}_R$ as well as $\mathbf{V}_I$ also can be obtained.

$$\begin{cases} \mathbf{U} = \mathbf{U}_R + i * \mathbf{U}_I \\ \mathbf{V} = \mathbf{V}_R + i * \mathbf{V}_I \end{cases} \quad (5.21)$$

The singular values of $\hat{\mathbf{Y}}_C$ can be determined by the diagonal values in matrix $\mathbf{W}$. In fact, the matrix can be expressed as following form:

$$\mathbf{W} = \begin{bmatrix} \delta_1 & 0 & \cdots & \cdots & 0 & 0 & 0 & \cdots & 0 \\ 0 & \delta_1 & \cdots & \cdots & 0 & 0 & 0 & \cdots & 0 \\ \vdots & \vdots & \ddots & \vdots & \vdots & \vdots & \vdots & \vdots & \vdots \\ \vdots & \vdots & & \ddots & \vdots & \vdots & \vdots & \vdots & \vdots \\ 0 & 0 & \cdots & \cdots & \delta_M & 0 & 0 & \cdots & 0 \\ 0 & 0 & \cdots & \cdots & 0 & \delta_M & 0 & \cdots & 0 \end{bmatrix} \quad \text{(for } M < N) \quad (5.22)$$

or

$$\mathbf{W} = \begin{bmatrix} \delta_1 & 0 & \cdots & \cdots & 0 & 0 \\ 0 & \delta_1 & \cdots & \cdots & 0 & 0 \\ \vdots & \vdots & \ddots & \vdots & \vdots & \vdots \\ \vdots & \vdots & \vdots & \ddots & \vdots & \vdots \\ 0 & 0 & \cdots & \cdots & \delta_N & 0 \\ 0 & 0 & \cdots & \cdots & 0 & \delta_N \\ 0 & 0 & \cdots & \cdots & 0 & 0 \\ \vdots & \vdots & \ddots & \ddots & \vdots & \vdots \\ 0 & 0 & \cdots & \cdots & 0 & 0 \end{bmatrix} \quad \text{(for } M > N) \quad (5.23)$$

As for the singular values in the matrix $\sum$, it has the form

$$\mathbf{\Sigma} = \begin{bmatrix} \delta_1 & 0 & \cdots & 0 & 0 & \cdots & 0 \\ 0 & \delta_2 & \cdots & 0 & 0 & \cdots & 0 \\ \vdots & \vdots & \ddots & \vdots & \vdots & \ddots & \vdots \\ 0 & 0 & \cdots & \delta_M & 0 & \cdots & 0 \end{bmatrix} \quad \text{(for } M < N) \quad (5.24)$$

or

$$\Sigma = \begin{bmatrix} \delta_1 & 0 & \cdots & 0 \\ 0 & \delta_2 & \cdots & 0 \\ \vdots & \vdots & \ddots & \vdots \\ 0 & 0 & \cdots & \delta_N \\ 0 & 0 & \cdots & 0 \\ \vdots & \vdots & \ddots & \vdots \\ 0 & 0 & \cdots & 0 \end{bmatrix} \text{ (for } M > N) \tag{5.25}$$

Form Eqs. (5.22) to (5.25), they give a relationship between the C-SVD and AMD. It means that the COMs and COVs of the complex ensemble measurement matrix also can be calculated by the augmented matrix based on its real and imagery parts.

5.3.4   Definition of Relevant COMs

A practical criterion for defining the approximate order of matrix can be obtained by the rational threshold. Generally speaking, the percentage energy contained in the relevant mode is given by

$$\sum_{i=1}^{p} \gamma_i \Big/ \sum_{i=1}^{N} \gamma_i = 99\ \% \tag{5.26}$$

That is to say

$$\sum_{1}^{p} \eta_i^2 / N \Big/ \sum_{1}^{N} \eta_i^2 / N = 99\ \% \tag{5.27}$$

p is the number of relevant modes.
The approximated error is described as:

$$\mathbf{E}_{\text{Error}} = \hat{\mathbf{Y}} - \text{Real}\left(\mathbf{U}_{(1:M,1:p)} \Sigma_{(1:p,1:p)} \mathbf{V}_{(1:N,1:p)}^H \right) \tag{5.28}$$

Actually, the decomposition time and accuracy can be adjusted by changing the number of the relevant COMs in COD.

5.4 Extraction of the Propagating Features

5.4.1 Spatial Energy Distribution

In [1], it has shown that the time-dependent complex coefficients associated with each eigenvector can be conveniently split into their amplitude and phase. From the COD analysis, the ensemble of data can be expressed as the complex expansion:

$$\hat{\mathbf{Y}}(y,t) = \sum_{i=1}^{p} \mathbf{R}_i(t) \angle \theta_i \mathbf{S}_i(t) \angle \varphi_i \tag{5.29}$$

where $\mathbf{R}_i(t)$ is the complex temporal amplitude function; $\mathbf{S}_i(t)$ is the complex spatial mode; θ_i is the temporal phase function; φ_i is the spatial phase function.

Obviously, these four parameters describe the basic propagation characteristics of the ith mode. Next, the mentioned parameters are calculated from temporal and spatial aspects. The similar idea can be seen in [1, 4]. However, the proposed approach in this section is different from the mentioned approaches in two areas. First, the temporal and spatial parameters are calculated separately in our methods while they are determined together in [1]. Furthermore, how to further apply the calculated parameters are not researched in mentioned publications. In this section, the physical interpretation of LFO is explained by following parameters.

5.4.2 Temporal Dynamic Characteristics

The spatial distribution characteristics of pth COMs can be analyzed from six aspects:

$$\begin{cases} \mathbf{B}_p(s) = \left(\mathbf{V}_{C_{(1:M,1:p)}} \times \Sigma^H_{C_{(1:p,1:p)}} \right) / \sqrt{M} \\ S_p(s) = \mathrm{Re}(\mathbf{B}_p(s)) \\ \phi_p(s) = \arctan(\mathrm{Im}[\mathbf{B}_p(s)] / \mathrm{Re}[\mathbf{B}_p(s)]) \\ Bm_p(s) = \mathrm{abs}(\mathbf{B}_p(s)) \\ a_p(s) = \phi_p(s) \times (180°/\pi) \\ k_p(s) = \mathrm{gradient}(\phi_p(s)) \end{cases} \tag{5.30}$$

where $\mathbf{B}(t)$ is the spatial matrix and satisfies the relation $\hat{\mathbf{Y}}_C = \mathbf{U}_C \times \mathbf{B}(s)^H$; $S_p(s)$ is the pth spatial distribution mode; $\phi_p(s)$ is the pth spatial phase function; $Bm_p(s)$ is the pth spatial amplitude function; $k_p(s)$ is the pth spatial distribution factor; $a_p(s)$ is the spatial angle function.

The temporal dynamic characteristics of pth COMs can be described by following parameters:

$$\begin{cases} \mathbf{A}_p(t) = \left(\mathbf{U}_{(1:M,1:p)} \times \Sigma_{(1:p,1:p)}\right)/\sqrt{N} \\ R_p(t) = \mathrm{Re}(\mathbf{A}_p(t)) \\ \theta_p(t) = \arctan(\mathrm{Im}[\mathbf{A}_p(t)]/\mathrm{Re}[\mathbf{A}_p(t)]) \\ Am_p(t) = \mathrm{abs}(\mathbf{A}_p(t)) \\ f_p(t) = \mathrm{gradient}(\theta_p(t))/2\pi \end{cases} \tag{5.31}$$

where $\mathbf{A}(t)$ is the temporal matrix and it satisfies the relation $\hat{\mathbf{Y}}_{\mathbf{C}} = \mathbf{A}(t) \times \mathbf{V}_{\mathbf{C}}^H$; $R_p(t)$ is the pth temporal oscillation mode; $\theta_p(t)\theta_p(t)$ is the pth temporal phase function; $Am_p(t)$ is the pth temporal amplitude function; $f_p(t)$ is the pth temporal oscillation frequency.

5.4.3 Energy Contribution Factor (ECF)

In order to compare with the NCF in Chap. 4, the concept of the energy contribution factor (ECF) is proposed in this subsection. As for the specific COM, ECF is defined as the ratio between energy of node (EON) and the maximal energy of COM (EOC) and it is used to validate energy contribution degree of each node to the specific COM. It is clear the ECF and the NCF which has been proposed in Chapt. 4 can be evaluated and compared each other.

$$\begin{cases} EON_i = S_p(i)(i \in [1,N]) \\ EOP_p = \sum_{i=1}^{N} \max(S_p(i)) \\ ECF_i = EON_i/EOP_p \end{cases} \tag{5.32}$$

5.5 The Flowchart of Proposed COD

From the description above and the knowledge in former chapters, the flowchart of proposed COD based on the ensemble measurement matrix is introduced as follows:

1. The measured signal obtained from WAMS is usually polluted by many kinds of noises. As for the single measured signal, improved EMD is used to determine the IMFs which contain the LFO components based on the energy analysis and frequency distribution. Therefore, the fluctuating part of the ensemble measurement matrix can be expressed as $\hat{\mathbf{Y}}$;

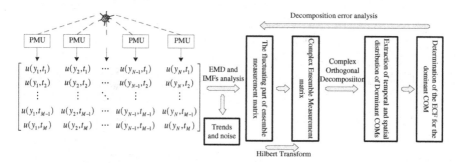

Fig. 5.3 The flowchart of the proposed COD

2. Each column of $\hat{\mathbf{Y}}$ is augmented to the form of complex analytical function by adding the imaginary part which is calculated by the HT. Then, the complex ensemble measurement matrix is built as $\hat{\mathbf{Y}}_{\mathbf{C}}$;

3. The C-SVD is employed to calculate the COMs, and then the temporal dynamic characteristics and spatial energy distributions of COMs are analyzed according to the energy relationship;

4. As for the dominant COM, the ECF of each measured signal is determined based on the spatial energy distribution.

5. The decomposition error of COD is analyzed.

The detailed flowchart of energy distribution analysis is shown in Fig. 5.3.

5.6 Study Case

5.6.1 Description of Sliding Window

The sliding window is a common tool for the analysis of measured signals. Here, the basic structure of the sliding window is described briefly.

Supposing that the length of time window is l and the size of time step is Δl, the data in window k can be expressed as y_{1-l}^{k}. Thus, the data in window $k + 1$ can be defined as:

$$y_{1-l}^{k+1} = y_{\Delta l + 1 - + \Delta l + l}^{k} \tag{5.33}$$

Obviously, the length of overlapped data between window k and window $k + 1$ is $l - \Delta l$. The schematic diagram of the sliding window is shown in Fig. 5.4.

Figure 5.4 shows that the data in a new window does not only depend on the former windows but also take new measured data into consideration. The sliding window gives a good combination between the former data and new data. It can be

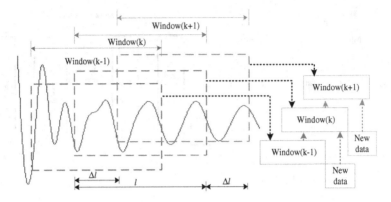

Fig. 5.4 Sliding time window

used to implement the near real-time application of the proposed ISM. In following sections, the COD-SWRA will be introduced, respectively. It should be noted that there is no essential difference between segmenting window and sliding window. As for the segmenting window, there is no overlapping in different windows. It means that the length of time window is equal to the step size. But the sliding window is defined as the step size which is shorter than the length of time window.

5.6.2   Sliding Window Recursive Algorithm (SWRA) of COD

For the specific time window, the ability of COD to extract the temporal variations and spatial distributions of COMs based on the ensemble matrix has been proved by the synthetic signal. In this section, the COD-SWRA will be described aiming at the near real-time application of the COD. The identification of the trend component from new measured data has been introduced in [5]. In order to extract the temporal and spatial parameter timely, the sliding detrend technique which includes the signal segment and moving average algorithm is proposed to determine the trend component in COD-SWRA.

Supposing the number of obtained signals is N, k is the sign of the time window and the step size is 1, in this window the covariance of extended ensemble measurement matrix can be written as the following form Eq. (5.35):

$$\begin{cases} \mathbf{F}_1^k = [\psi(y_{1:N}, t_{1:k})]^H [\psi(y_{1:N}, t_{1:k})] \big/ k \\ \mathbf{F}_2^k = [\psi(y_{1:N}, t_{1:k})]^H [\psi(y_{1:N}, t_{1:k})] \big/ N \end{cases} \qquad (5.34)$$

In the next sampling point, the covariance of new matrix are extended as

$$
\begin{cases}
\mathbf{F}_1^{k+1} = [\psi(y_{1:N}, t_{1:k+1})]^{\mathbf{H}}[\psi(y_{1:N}, t_{1:k+1})]\Big/ k \\
\quad = \frac{[\psi(y_{1:N},t_{1:k})]^{\mathbf{H}}[\psi(y_{1:N},t_{1:k})]}{k} + \frac{[\psi(y_{1:N},t_{k+1})]^{\mathbf{H}}[\psi(y_{1:N},t_{k+1})]}{k} \\
\mathbf{F}_2^{k+1} = [\psi(y_{1:N}, t_{1:k+1})]^{\mathbf{H}}[\psi(y_{1:N}, t_{1:k+1})]\Big/ N \\
\quad = \frac{[\psi(y_{1:N},t_{1:k})]^{\mathbf{H}}[\psi(y_{1:N},t_{1:k})]}{N} + \frac{[\psi(y_{1:N},t_{k+1})]^{\mathbf{H}}[\psi(y_{1:N},t_{k+1})]}{N}
\end{cases}
\tag{5.35}
$$

Substituting Eq. (5.35) into Eq.(5.36), the following equations can be obtained

$$
\begin{cases}
\mathbf{F}_1^{k+1} = \mathbf{F}_1^{k} + \frac{[\psi(y_{1:N},t_{k+1})]^{\mathbf{H}}[\psi(y_{1:N},t_{k+1})]}{k} \\
\mathbf{F}_2^{k+1} = \mathbf{F}_2^{k} + \frac{[\psi(y_{1:N},t_{k+1})]^{\mathbf{H}}[\psi(y_{1:N},t_{k+1})]}{N}
\end{cases}
\tag{5.36}
$$

This analysis suggests that the covariance matrix at time $k + 1$ does not only depend on the previous covariance matrix at time k, but also depends on the current measurements at the wide area. That provides a recursive algorithm for the near real-time application of the C-SVD.

Similarly, the length of time window and size of the time step can be selected adaptively. The Eq. (5.37) can be extended to following forms:

$$
\begin{cases}
\mathbf{F}_1^{k+1} = \frac{l-\Delta l}{l} \mathbf{F}_1^{k} + \frac{[\psi(y_{1:N},t_{(l-\Delta l):l})]^{\mathbf{H}}[\psi(y_{1:N},t_{(l-\Delta l):l})]}{l} \\
\mathbf{F}_2^{k+1} = \frac{N-\Delta l}{N} \mathbf{F}_2^{k} + \frac{[\psi(y_{1:N},t_{(l-\Delta l):l})]^{\mathbf{H}}[\psi(y_{1:N},t_{(l-\Delta l):l})]}{N}
\end{cases}
\tag{5.37}
$$

According to the description in this chapter, the spatial and temporal parameters of the dominant COM can be extracted by C-SVD if the two forms of covariance matrixes are determined. The flowchart of the proposed recursive algorithm is shown in Fig. 5.5.

5.6.3 Applications of the COD-SWRA

As noted earlier and the theory in [6], the COD-SWRA will be used to extract the temporal variation and spatial distribution of the dominant COM. The length of time window is set as 30 s and the step size is selected as 10 s. The oscillation mode at 23:57:00:00 is also set as the research subject form two aspects: one considers the data as the new measured data (Data range is from 56:35:01 to 57:05:00); the other selects the data in the former time window as previous data (Data range is from 56:45:01 to 57:15:00). Obviously, these two are continuous sliding windows.

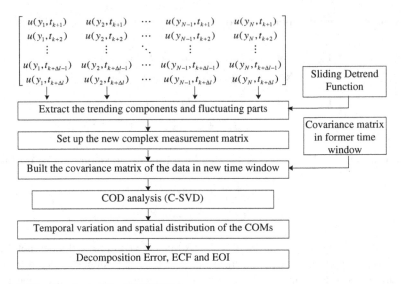

Fig. 5.5 The flowchart of the COD-SWRA

The data in the sliding time windows are shown in Fig. 5.6. The red dot–dashed line indicates the first time window and the blue dot–dashed line mark the following time window.

After the detrending and the construction of the complex ensemble measurement matrix, the C-SVD is utilized to extract the COMs according to the energy relationships. Furthermore, the number of relevant COMs is determined by the decomposition error. The energy ratios of the COMs (decomposition error is 1 %) in the first and the second time window are shown in Fig. 5.7.

In Fig. 5.7, the left picture means the energy ratios of the first three COMs in first time window. The energy ratio of COM1 is about 83.04 % and the second COM is close to 15.17 %. The right picture stands for the energy ratios of the first three

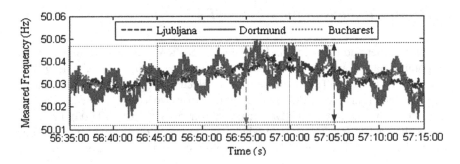

Fig. 5.6 The data in the sliding time windows

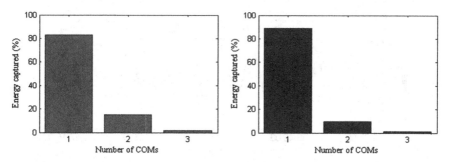

Fig. 5.7 Energy ratios of COMs in the first and second windows

COMs in the second time window. It is clear that the energy ratio of COM1 is changed from 83.04 to 88.93 % while the energy ratio of the COM2 is deduced from 15.17 to 9.76 %. It means that the energy of second COM is transferred to the COM1.

Next, the temporal variations of the dominant COM in the first and second time window are shown in Fig. 5.8.

The calculated results are analyzed from four aspects: (1) the overlapped COM1 which belongs to different time window fit very well. It means that the proposed COD-SWRA can extract the dominant COM in an accurate way; (2) the calculated frequency is time-varying and nonlinear. However, the values are fluctuating with the 0.25 Hz; (3) for the calculated amplitude, it can be seen that there are obvious increasing tendency both in the first and the second time window. This phenomenon can be verified from the energy ratios of the dominant COMs in Fig. 5.7; (4) the temporal phases of the COM1 in the two time windows are nearly constant.

From the temporal parameters of the COM1, it can be estimated that there is an increased oscillation mode and the frequency of this oscillation mode is near 0.25 Hz. However, the spatial information of the dominant COM is unknown. According to the calculation formulas in Chapter 5, the spatial distribution parameters of the COM1 are shown in Fig. 5.8. Moreover, the ECFs of each measured signal in different time window are determined in the Table 5.1.

Table 5.1 shows that the changes of the ECFs in the first and second time window. The dominant COM is an increasing oscillation mode. At the same time, the distribution of its energy is also time-varying. The ECF of Ljubljana is decreasing while the ECF of Bucharest is increasing. It means that the contribution of the Bucharest to the COM1 has a rising trend.

Besides, the mode shapes of the dominant COMs in different time window are determined respectively. And the results are shown in Fig. 5.9.

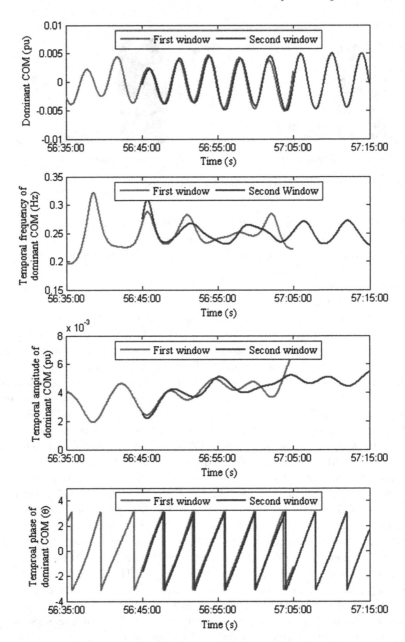

Fig. 5.8 The temporal variation of COM1 in the first and second windows

Table 5.1 The ECFs in COM1 in the first and second windows		Ljubljana	Dortmund	Bucharest
	First window	−0.1349	1	0.2920
	Second window	−0.1258	1	0.3052

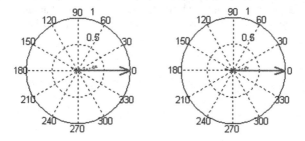

Fig. 5.9 Mode shape of the COM1 in different time window

5.7 Summary

In this chapter, from the viewpoint of oscillation energy distribution in wide area, temporal, and spatial characteristics of oscillation modes in power system are analyzed based on the ensemble measurement matrix. COD provides an efficient and accurate approach to explore the temporal variation and spatial distribution of oscillation processes without a priori assumption. These parameters are important in determining the control strategies and protect device actions. In order to realize the near real-time application, three different COD methods, including the C-ED, C-SVD, and AMD, are compared under the different sizes of ensemble measurement matrixes. Results of numerical simulation show that the C-SVD has obvious advantage in calculation speed.

The application of the actual measured data from WAProtector has been used to validate the performance of the proposed COD-SWRA method. The results reveal that the COD-SWRA can be easily implemented to extract the dominant oscillation mode near real-time and their calculated results are consistent. At the same time, the COD is utilized to extract the temporal variation and spatial distribution of the dominant COM in the wide area and to provide control strategy to the transmission system operators.

References

1. Esquivel P (2010) Extraction of dynamic patterns from wide area measurements using empirical orthogonal functions. A dissertation for PhD degree, CINVESTAV del IPN Unidad Guadalajara, Aug 2010
2. Yang QX, Bi TS, Wu JT (2007) WAMS implementation in China and the challenges for bulk power system protection. IEEE Power and Energy Society General Meeting, 24–28 June 2007

3. Yang DC, Rehtanz C, Li Y et al (2012) Denoising and detrending of measured oscillatory signal in power system. Electr Rev 88(3b):135–139
4. Messina AR (2009) Inter-area oscillations in power systems: a nonlinear and non-stationary perspective. Springer, Berlin. ISBN 978-0-387-89529-1
5. Messina AR, Vittal V, Heydt GT et al (2009) Non-stationary approaches to trend identification and denoising of measured power system oscillation. IEEE Trans Power syst 24(4):1798–1807
6. Messina AR, Esquivel P, Lezama F (2010) Time-dependent statistical analysis of wide-area time-synchronized data. Math Probl Eng. Article ID 751659, 17 p. doi:10.1155/2010/751659

Chapter 6
Basic Framework and Operating Principle of Wide-Area Damping Control

With the technical development and cost reduction of power electronics, more and more FACTS devices are being applied into power system especial into modern power system toward smart grid [1–3]. FACTS devices can conveniently adjust network parameters (e.g., reactance) flexibly and benefit power flow optimization, transmitted power increase, bus voltage stability enhancement, and so on. At the same time, with more and more application of synchronized phasor measurement (PMU) technology, WAMS technology are being applied into power system, which is also one obvious technical feature of the coming smart grid [4–6]. Therefore, it would be wonderful to construct WAMS-based FACTS supplementary wide-area damping control strategy, that combines the quick and flexible control ability of FACTS devices and the global monitoring ability of WAMS, to prevent the low-frequency oscillation (especial the inter-area oscillation) and enhance the global stability of power system.

In this chapter, the basic framework of wide-area damping control will be represented, and the operating principle that using the supplementary control function of FACTS device to construct the wide-area damping control strategy will be investigated in detail. The delay effect of wide-area control signal on the damping performance is also discussed. A case study will be performed to represent the basic system modeling method, the advanced robust control design method, and the basic damping performance that handles with the time delay of wide-area signal.

6.1 Basic Framework of Wide-Area Damping Control

The application of wide-area measurement system (WAMS) technology in power system makes it convenient to monitor and transmit system operating variables within global range. Generally, by way of global position system (GPS), optical fiber communication, or other advanced communication technologies, WAMS can perform synchronized measurement and centralized processing of remote phasor data. Furthermore, to enhance system stability, optimize power flow, increase protection ability, or other positive purpose for operating improvement of power system, it could be possible to utilize the WAMS technology and select remote

© Springer-Verlag Berlin Heidelberg 2016
Y. Li et al., *Interconnected Power Systems*, Power Systems,
DOI 10.1007/978-3-662-48627-6_6

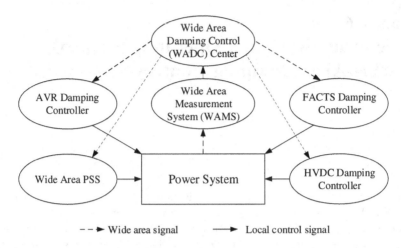

Fig. 6.1 Basic framework of wide-area damping control

signal as the feedback-control input to implement various wide-area control and protection strategies. In this book, the wide-area damping control is concerned deeply for stability enhancement of power system. Figure 6.1 shows the basic framework that utilizes the main control devices and WAMS to implement wide-area damping control strategy.

According to Fig. 6.1, it can be seen that the framework of wide-area damping control basically includes control devices, WAMS, WADC, and power system. The WAMS is used to synchronously measure and process the global operating variables; and at the same time, these processed variables are then submitted to wide-area damping control (WADC) center. The WADC center is used to select the suitable control-input signals, and then through the centralized wide-area damping controller, it sends the wide-area feedback-control signals to the local control devices. The local control devices, such as FACTS controller, AVR, PSS, HVDC, and so on, receive the corresponding wide-area feedback signals and implement the separate supplementary damping control. In the control framework, WADC is a very important part, which mainly includes the choice of wide-area control-input signals and the generation of wide-area control-output signals. For the WADC design, nowadays, there are various design methods that can be utilized to guide the controller design, such as the classic phase-compensation method [7, 8], the linear control design method [9], the robust design method [10, 11], and so on.

Furthermore, it should be noted that during the transmission and processing of wide-area signals, there inevitably exists time delays that endanger the practical damping performance. From Fig. 6.1, it can be found that the transmission delay mainly includes the following two parts: (1) when the measured signals are transmitted from different areas of power system to WAMS center, the time delays must be generated. (2) Similarly, when the wide-area control signals are transmitted from WADC center to local control devices, another time delays are also generated. Besides these, for the wide-area signals in both WAMS and WADC center, the signal process inevitably wastes time which also leads to a certain amount of time delays. Many research results show that the time delays even a little time delays may worse or even fail the control performance. However, for the controller design, the conventional design methods mentioned in the above section cannot consider or handle with the delay effect very well. In this book, the delay effect is paid the deep attention, and multitype advanced controller design methods will be proposed to decrease or eliminate the effect of time delay on the wide-area damping performance, which will be discussed in the following chapters.

Moreover, as Fig. 6.1 shows, various power system control devices can be used to introduce wide-area supplementary control-input signal to implement oscillation damping control, such as power system stabilizer (PSS), automatic voltage regulator (AVR), control devices of flexible ac transmission system (FACTS), high-voltage direct current (HVDC) transmission system, and so on. In practice, the choice of wide-area control device should comprehensively consider the practical control performance, the investment cost, and other potential factors. In this book, the FACTS devices are used to implement wide-area damping control strategy. Because that compare with PSS and AVR, FACTS devices are convenient to perform centralized control, and they can also play effective damping role on critical inter-area oscillation modes. For AVR and PSS, both of them are associated to generator located in different area of power system, and they are usually designed to damp local oscillation modes related to their associated generators. In such a case, if they are used to perform wide-area control strategy, the coordination of various AVR or PSS associated to various generators, and the damping effect on both local and inter-area oscillation modes, should be considered carefully. In addition, with more and more application of FACTS devices in power system, it changes more and more convenient to use their supplementary control functions and implement wide-area damping on inter-area oscillation among different areas; and at the same time, PSS and AVR can keep on effective damping on local oscillation related to their associated generators.

6.2 Operating Principle of Wide-Area Damping Control

In essence, the purpose of FACTS wide-area damping control is to shift all the system eigenvalues into the left-hand side of the S-plane, so that the needed minimum damping ratio can be satisfied and the system damping can be increased into the acceptable level. The operating principle on system damping increase by means of FACTS wide-area damping control can be revealed through the simplified wiring diagram of single-machine infinite-bus (SMIB) system installed with FACTS

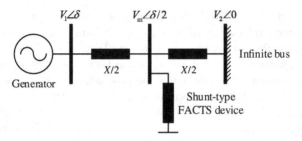

Fig. 6.2 Simplified wiring diagram of single-machine infinite-bus (*SMIB*) system with FACTS device

device, as shown in Fig. 6.2. In this SMIB system, one shunt-type FACTS device (e.g., SVC, STATCOM) is installed at the middle of the transmission line.

According to Fig. 6.2, the magnitude value of the buses can be expressed as

$$\begin{cases} V_1 = |V_1| \sin(\omega t + \delta) \\ V_2 = |V_2| \sin \omega t \\ V_m = |V_m| \sin(\omega t + \delta/2) \end{cases} \tag{6.1}$$

The electromagnetic power of the generator can be written as

$$P_E = \frac{2VV_m}{X} \sin\frac{\delta}{2} \tag{6.2}$$

The increment equation of the generator electrical power can be obtained by linearizing the above equation, that is,

$$\Delta P_E = \frac{\partial P_E}{\partial V} \Delta V + \frac{\partial P_E}{\partial V_m} \Delta V_m + \frac{\partial P_E}{\partial \delta} \Delta \delta \tag{6.3}$$

Assume that when the bus voltage associated to the generator keeps on constant and then combine the rotor motion equation of generator, the above equation can be further expressed as

$$M\frac{\mathrm{d}^2(\Delta\delta)}{\mathrm{d}t^2} + \frac{\partial P_E}{\partial V_m} \Delta V_m + \frac{\partial P_E}{\partial \delta} \Delta \delta = 0 \tag{6.4}$$

Because that the installed FACTS device can control the voltage of the associated bus, if the bus voltage is controlled as the following functional relationship:

$$\Delta V_m = K\frac{\mathrm{d}(\Delta\delta)}{\mathrm{d}t} \tag{6.5}$$

where K is a constant value.

Then the increment (6.4) can be further rewritten as

$$M\frac{d^2(\Delta\delta)}{dt^2} + \frac{\partial P_E}{\partial V_m}\bigg|_0 K\frac{d(\Delta\delta)}{dt} + \frac{\partial P_E}{\partial \delta}\bigg|_0 \Delta\delta = 0 \tag{6.6}$$

According to the above equation, the system damping can be represented as

$$2\zeta = \frac{K}{M}\frac{\partial P_E}{\partial V_m}a \tag{6.7}$$

Beside the voltage stability control that FACTS device can benefit for power system, from the above equation, it can be also seen that the application of FACTS device can also provide certain positive damping, and such technical feature is advantage to develop oscillation damping control through introducing supplementary wide-area damping signal for FACTS controller. Figure 6.3 represents the general configuration.

As Fig. 6.3 shows, the wide-area damping controller (WADC) is in fact the supplementary controller of local FACTS controller, and there is the supplementary wide-area control-output as one part of the control-input of the local FACTS controller. The WAMS is in charge of collection, transmission, and processing wide-area signals through a cluster of PMUs located in different areas of power system. Combining Fig. 6.1 and the related analysis, the WAMS center sends the processed wide-area signals to WADC center. Following this, the WADC center further analyzes these wide-area signals and selects the suitable wide-area signal as the control-input of WADC. The WADC is in charge of generating an effective wide-area control-output that can provide enhanced damping through the simultaneous action of local FACTS controller and WADC on the power system with FACTS device.

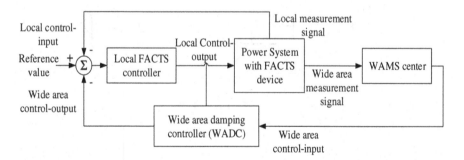

Fig. 6.3 General configuration of FACTS wide-area damping control

6.3   System Modeling

6.3.1   SMIB System with FACTS WADC

The single line diagram of SMIB system with shunt-type FACTS device (SVC) is shown more in detail in Fig. 6.4. As Fig. 6.4 shows, there are two kinds of control-input signals for the SVC controller with WADC. One is the wide-area signal transmitted from the remote generator, that is the rotor angle in electrical degree δ and angular velocity ω of the remote generator, and another is the local signal measured from the bus shunted with SVC device, that is, the bus voltage V_s. The control target of the SVC controller with WADC is to damp the system power oscillation through wide-area damping control, and in the meantime to maintain the stability of the local bus voltage.

6.3.2   System Modeling Based on Direct Feedback Linearization Theory

For the controller design, the system model that can reflect the small signal stability of the test system should be obtained in advance. In this chapter, the direct feedback linearization (DFL) theory [12, 13] is presented to perform the system modeling. According to the electrical connection diagram shown in Fig. 6.4, the equivalent circuit diagram of the SMIB system can be established as shown in Fig. 6.5, in which, E expresses the transient voltage of the generator; x_1 and x_2 express the equivalent reactance of the sending and the receiving end, respectively; and

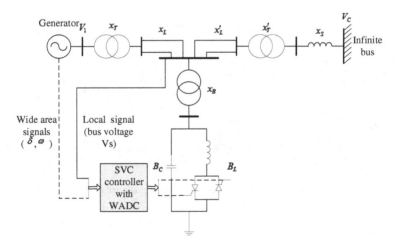

Fig. 6.4 Single line diagram of SMIB system with SVC device

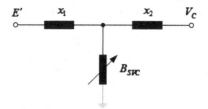

Fig. 6.5 Equivalent circuit diagram of SMIB system with SVC device

$x_1 = x_d' + x_T + x_L$, $x_2 = x_L' + x_T' + x_S$, where x_d' expresses the d-axis transient reactance of the equivalent generator, x_T and x_T' express the equivalent leakage reactance of the transformers of the sending and the receiving end, respectively; x_L and x_L' express the equivalent reactance of the transmission line divided by the shunted SVC device, and x_S expresses the system equivalent reactance; B_{SVC} expresses the equivalent susceptance of the SVC, and $B_{SVC} = B_C + B_L$, where, B_C and B_L express the equivalent susceptance of the mechanically switched capacitor (MSC) and the thyristor controlled reactor (TCR), respectively.

Assume that the generator is expressed by second-order model, that is, E' is kept constant, and the mechanical power P_m of the generator is also kept constant, then, the state equation of the above system can be obtained as follows:

$$\begin{cases} \Delta\dot{\delta} = \omega - \omega_0 \\ \Delta\dot{\omega} = -D(\omega - \omega_0)/H + \omega_0(P_m - P_e)/H \\ \Delta\dot{B}_L = (-B_{SVC} + B_{SVC_0} + K_C u_B)/T_C \end{cases} \tag{6.8}$$

where

$$P_e = \frac{E'V_C}{x_1 + x_2 + x_1 x_2 B_{SVC}} \sin\delta \tag{6.9}$$

in which, ω_0 expresses the synchronous speed, and $\omega_0 = 2\pi f_0$; P_e expresses the electromagnetic power; D and H express the damping coefficient and the inertia time constant of the generator, respectively; K_C and T_C express the magnification times and the inertia time constant of the adjustable system of the SVC, respectively; V_C expresses the infinite-bus voltage; and the other parameters are as the mentioned above.

In order to ensure the voltage stability of the bus shunted with SVC device, the additional state variable about the local bus voltage should be introduced according to internal model control, that is,

$$\Delta V_S = \int (V_S - V_{S_0}) dt \tag{6.10}$$

The additional voltage state variable can be expressed as

$$\Delta \dot{V}_S = V_S - V_{S_0} \tag{6.11}$$

where

$$V_S = \frac{\sqrt{(x_2 E')^2 + (x_1 V_C)^2 + 2x_1 x_2 V_C E'C \cos \delta}}{x_1 + x_2 + x_1 x_2 B_{SVC}} \tag{6.12}$$

Combining the above equation with (6.9), the following equation can be obtained:

$$V_S = \frac{\sqrt{(x_2 E')^2 + (x_1 V_C)^2 + 2x_1 x_2 V_C E'C \cos \delta}}{E' V_C \sin \delta} P_e \tag{6.13}$$

Because the voltage control could be considered near the operational point, $\Delta \dot{V}_S$ can be further obtained by the method of Taylor Series Expansion:

$$\Delta \dot{V}_S = \frac{\partial V_S}{\partial \delta} \Delta \delta|_0 + \frac{\partial V_S}{\partial P_e} \Delta P_e|_0 = S_\delta \Delta \delta + S_{P_e} \Delta P_e \tag{6.14}$$

If we select the following equation as the virtual controllers:

$$\Delta \dot{P}_e = P_m - P_e = P_m - \frac{E' V_C}{x_1 + x_2 + x_1 x_2 B_{SVC}} \sin \delta \tag{6.15}$$

Then, following the direct feedback linearization (DFL) theory, by combining (6.8) with the above (6.14) and (6.15), the initial linearized state equation can be obtained as:

$$\dot{x}(t) = Ax(t) + B_1 u(t) \tag{6.16}$$

where

$$A = \begin{bmatrix} 0 & \omega_0 & 0 & 0 \\ 0 & -\frac{D}{H} & \frac{\omega_0}{H} & 0 \\ 0 & 0 & 0 & 0 \\ S_\delta & 0 & S_{P_e} & 0 \end{bmatrix}, \quad B_1 = \begin{bmatrix} 0 \\ 0 \\ 1 \\ 0 \end{bmatrix}.$$

The above equation is the linearized model of SMIB system with SVC device, which can be used for the robust linear design of SVC local controller with WADC.

6.4 Summary

In this chapter, the overall framework of wide-area damping control is presented and analyzed briefly. Based on such framework, the control concept that utilizes the flexible and quick control feature of FACTS device to implement the WAMS-based wide-area damping control is described in detail. The operating principle of FACTS wide-area damping control is presented by studying a SMIB system with shunt-type FACTS device, which indicates that FACTS device can not only achieve local control strategy (e.g., bus voltage stability, power flow control) but also realize the oscillation damping through wide-area control strategy. Such damping performance is achieved by enhancing system damping ratio through introducing the suitable wide-area control-input. Furthermore, the general configuration of FACTS wide-area damping control is presented, which is advantage to guide the control design represented in the following chapters. Finally, the system linearized modeling method (direct feedback linearization, DFL) is described.

References

1. Bose A (2010) New smart grid applications for power system operations. In: IEEE power and energy society general meeting, 2010
2. Zarghami M, Crow ML, Jagannathan S (2010) Nonlinear control of FACTS controllers for damping interarea oscillations in power systems. IEEE Trans Power Delivery 25(4): 3113–3123
3. Eriksson R, Soder L (2011) Wide-area measurement system-based subspace identification for obtaining linear models to centrally coordinate controllable devices. IEEE Trans Power Delivery 26(2):988–997
4. Bose A (2010) Smart transmission grid applications and their supporting infrastructure. IEEE Trans Smart Grid 1(1):11–19
5. De La Ree J, Centeno V, Thorp JS, Phadke AG (2010) Synchronized phasor measurement applications in power systems. IEEE Trans Smart Grid 1(1):20–27
6. Chakrabortty A, Chow JH, Salazar A (2011) A measurement-based framework for dynamic equivalencing of large power systems using wide-area phasor measurements. IEEE Trans Smart Grid 2(1):68–81
7. Demello FP, Concordia C (1969) Concepts of synchronous machine stability as affected by excitation control. IEEE Trans Power Apparatus Syst 88(4):316–329
8. Cai LJ, Erilich I (2005) Simultaneous coordinated tuning of PSS and FACTS damping controllers in large power systems. IEEE Trans Power Syst 20(1):294–300
9. Gupta R, Bandyopadhyay B, Kulkarni AM (2005) Power system stabiliser for multimachine power system using robust decentralised periodic output feedback. IEE Proc Control Theory Appl 152(1):3–8
10. Zhang Y, Bose A (2008) Design of wide-area damping controllers for interarea oscillations. IEEE Trans Power Syst 23(3):1136–1143
11. Kutzner R, Scholz B, Reimann M (2003) An advanced model-based approach to stabilize power system oscillations based on the H-infinite theory modelling and tuning guide, practical experience. In: IEEE power engineering society general meeting, 2003

12. Zhu CL, Zhou RJ, Wang YY (1998) A new decentralized nonlinear voltage controller for multimachine power systems. IEEE Trans Power Syst 13(1):211–216
13. Grcar SV, Kumar SVJ, Yadaiah N (2004) Control strategies for transient stability of multimachine power systems-a comparison. In: IEEE international conference on power system technology, 2004

Chapter 7
Coordinated Design of Local PSSs and Wide-Area Damping Controller

The topic on how to coordinate local and wide-area damping control is investigated in this chapter. A sequential design and global optimization (SDGO) method is proposed to coordinate local PSSs and HVDC-WADC simultaneously. The sequential design (SD) is used to determine the time constants of phase-compensation blocks of PSS and HVDC-WADC. Afterward, the global optimization (GO) issued to optimize their control gains. The proposed method focuses on damping all dominant modes including local and inter-area modes to enhance the overall stability of AC/DC interconnected systems.

7.1 Overview of Optimization Method

The WADC can enhance the stability of interconnected systems, but it is usually used to damp the inter-area mode. In practice, the local damping control should still be involved to coordinate WADC for the overall stability enhancement. Normally, the HVDC-WADC is used to damp the inter-area oscillation mode, and the PSS is used to damp the local oscillation mode. In this situation, the controller interaction between PSS and HVDC-WADC should be considered carefully. In practical interconnected systems, many PSSs are installed in different areas of the network. To avoid the interaction between local and wide-area control, and improve the overall stability, it is important to tune PSSs and HVDC-WADC simultaneously.

So far various simultaneous tuning methods have been proposed to optimize PSS or other power system damping controllers [1–5]. These methods can be summarized as follows:

- These methods only tune the local damping controllers, and do not consider the interaction and coordination between local and wide-area damping controllers.
- Most of them need to optimize the control gain and the phase-compensation block simultaneously. Taking the 16-machine 5-area interconnected systems as an example [3, 6], for a PSS unit, there are at least three parameters (i.e., one control gain and two lead-lag time constants) that should be optimized. For the whole systems, at least $16 \times 3 = 48$ parameters have to be optimized

© Springer-Verlag Berlin Heidelberg 2016
Y. Li et al., *Interconnected Power Systems*, Power Systems,
DOI 10.1007/978-3-662-48627-6_7

simultaneously. On this condition, it has to consume much time to optimize the parameter set, which is uneconomic to the application in practical power systems.

- Most of them enhance the stability via the coordination control of all the PSSs. Sometimes, if parts of PSSs fail the damping function, it may affect the control performance of other PSSs.
- The objective function of these methods usually focus on all oscillation modes including the stable modes (damping ratio is more than 0.05), which inevitably increases the computational burden in each iteration step.

7.2 Description of Sequence Design and Global Optimization Method

7.2.1 Structure of PSS and HVDC-WADC

Figure 7.1 shows the controller structure of the PSS and the HVDC-WADC connected with the static exciter and the HVDC converter pole-controller, respectively. By the auxiliary control on the excitation system, PSS produces a component of electrical torque in phase with the rotor speed deviations [7]. It generally contains one control gain, one washout block and two phase-compensation blocks. The HVDC-WADC adopts the similar structure, but the difference is that its control-input is selected not only from the local signals but also from the wide-area signals. The study on the signal selection is presented in Chap. 9.

Note that for the HVDC system, the pole-controller at the rectifier or the inverter side can be added the supplementary WADC. Here, considering that the pole converter at the rectifier side has a simple controller structure, so the WADC is implemented at this side.

7.2.2 Design Procedure

Figure 7.2 shows the flowchart of SDGO method. The procedure of the simultaneously tuning of PSSs and HVDC-WADC is explained as follows:

- *Step-1*: the AC/DC interconnected systems are linearized at the normal operating condition. Nowadays, there are various modeling methods supporting the system linearization [7–14]. Especially for the smart transmission grid with advanced wide-area measurement, the transitional real-time modeling system and database [10] can be used for the system linearized modeling.

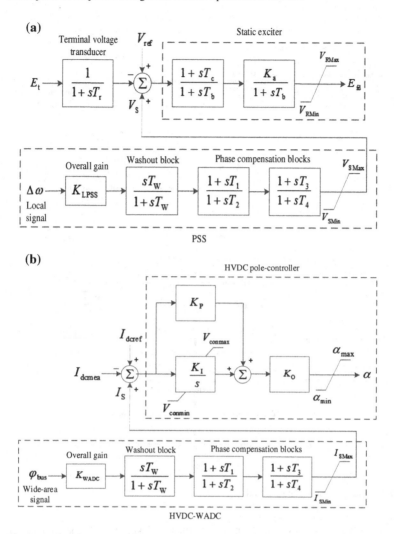

Fig. 7.1 Controller structure, (**a**) PSS linked with the static exciter; (**b**) HVDC-WADC linked with the converter pole-controller

- *Step-2*: the modal analysis is used to obtain the dominant oscillation modes. In this step, the small signal stability analysis (SSSA) [7, 10] or the real-time state estimation method [11, 15, 16] can be used for modal analysis.
- *Step-3*: all dominant modes are, respectively, distributed to each PSS or HVDC-WADC. Based on the dominant modes and the generator participation factors obtained by the modal analysis above, we can distribute the dominant modes to each PSS. Meanwhile, considering the high damping ability of the HVDC-WADC on the inter-area mode, so the typical inter-area mode is assigned to it.

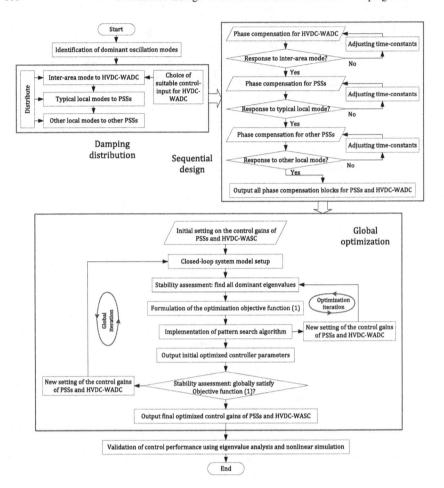

Fig. 7.2 Flowchart of SDGO method

- *Step-4*: the sequential design for each PSS and HVDC-WADC is carried out. The purpose of sequential design is to configure each dominant mode on each LPSS and HVDC-WADC by designing phase-compensation blocks step-by-step. For the design method, the angle-lag matching [12] or compensation techniques [13] can be adopted. Besides, in this step a suitable control-input is selected for the phase-compensation design of HVDC-WADC.
- *Step-5*: the global optimization is carried out to find optimal control gains for all PSSs and HVDC-WADC. In this step, the objective function is established to ensure that all the dominant modes are shifted into the acceptable region (damping ratio is more than 0.05) in the *S*-plane. The pattern search algorithm is used to solve the objective function.

- *Step-6*: the operating performance of the interconnected systems installed the optimized PSSs and HVDC-WADC should be validated under various operating conditions. Here, the eigenvalue analysis and nonlinear simulation are used for the validation.

7.3 Methodological Implementation

7.3.1 Damping Distribution

The damping distribution is in charge of selecting a suitable PSS or HVDC-WADC to damp the considered dominant mode. It mainly includes three steps, i.e., *Step-1*: Distribution of the inter-area mode to HVDC-WADC, *Step-2*: Distribution of the typical local modes to PSSs, and *Step-3*: Distribution of other LFO modes to other PSSs. These steps will be explained more detailed in Sect. 7.4.2.

7.3.2 Sequential Design

The objective of sequential design (SD) is to get the lead-lag time constants of the phase-compensation blocks of each PSS and HVDC-WADC step-by-step, so that each controller is able to damp the assigned dominant mode. To realize this objective, a phase-matching [12] and a phase-compensation [13] methods are adopted for PSS and HVDC-WADC. During each step of SD, the root-locus analysis is carried out to check the effectiveness of the designed phase-compensation blocks on the assigned dominant mode. The guideline of SD is made as follows:

1. For the HVDC-WADC, initially design the lead-lag blocks for phase-compensation at the frequency of inter-area mode. Then, the compensation performance by means of root-locus analysis is checked. If there is not effective performance, adjust the lead-lag time constants until the acceptable effectiveness achieves.
2. For the PSSs assigned to damp the typical local modes, in each step, initially design the lead-lag blocks to match the phase lag between the exciter voltage reference and the generator electric power output. Then, the damping performance is evaluated by means of root-locus analysis. If the effective performance can be obtained, then assess the performance on other dominant modes. Otherwise, if there is no performance, adjust the parameters of the lead-lag blocks until the acceptable performance can be reached, afterward the damping performance on other modes is evaluated.

3. For the PSSs assigned to damp the other local modes, in each step, first, the damping performance of the designed PSSs in (2) on the local modes concerned in (3) is evaluated. If these PSSs cannot provide a sufficient damping on these modes, then, initially design the lead-lag blocks of the PSS installed at the generator with the highest participation factor, and then evaluate the damping performance by means of root-locus analysis. Otherwise, if these PSSs can provide effective damping, the PSS related to the generator with the second highest participation factor is considered to damp these dominant modes.

7.3.3   Global Optimization

The global optimization (GO) method based on a nonlinear optimization program is proposed to optimize the control gains of PSSs and HVDC-WADC. The optimization object is to reconfigure the eigenvalues globally and ensure that the damping ratio of each dominant mode is increased from less than 0.05 to more than 0.05. The objective function is expressed as follows:

$$
\min. \quad F = \begin{cases} \text{if existing } \varsigma_i < 0.05, & \min\left(\sum_{i=1}^{n} (1 - \varsigma_i) \right)_j \\ \text{else,} & 0.95 \end{cases} \quad (7.1)
$$

$$
\text{subject to } 0 \le K_{\text{PSS and HVDC-WADC}} \le 100
$$

where F is the function value whose final optimal value is defined as 0.95; ς_i is the damping ratio of the i-th oscillation mode; n is the number of the dominant mode with damping ratio less than 0.05; j is the number of the considered operating condition; and K is the control gain of PSSs and HVDC-WADC.

Unlike the objective functions proposed in [1–5], the number n in (7.1) is decreased gradually during the iteration process, so the objective function represents an adaptive characteristic, which can reduce the computational cost and accelerate the convergence of the optimization algorithm.

The flowchart of GO is shown in Fig. 7.2. This method uses the pattern search algorithm to calculate (7.1). It mainly includes the following aspects:

- *Stability assessment*: Before the formulation of (7.1), the stability assessment is implemented to find all dominant modes with damping ratio less than 0.05, and offer them to (7.1). After the output of the optimized control gains, the stability assessment is implemented again to identify whether there still exists unstable modes or not.
- *Optimization iteration*: This kind of iteration is based on the pattern search algorithm. In each iteration step, the former generated control gains are used to set up the later closed-loop system. After the stability assessment, the (7.1) newly formed is supplied for the implementation of pattern search algorithm.

- *Global iteration*: This kind of iteration is the complementarity of the optimization iteration. Sometimes, the optimization iteration falls into an infinite loop and cannot obtain the acceptable result. In this situation, the resetting on the initial control gains is carried out, and then the global iteration can be implemented to restart the new procedure of the optimization iteration.

7.4 Case Study

7.4.1 AC/DC Hybrid Interconnected Systems

The 16-machine 5-area interconnected systems shown in Fig. 7.3 are used to test the SDGO method. In fact, it is a typical interconnected system including the New England Test System (NETS) and the New York Power System (NYPS). It is divided into five areas. G1–G9 and G10–G13 belong to Area-5 and -4, respectively, and the other three machines G14–G16 are equivalent to three neighbor areas interconnecting Area-4 separately.

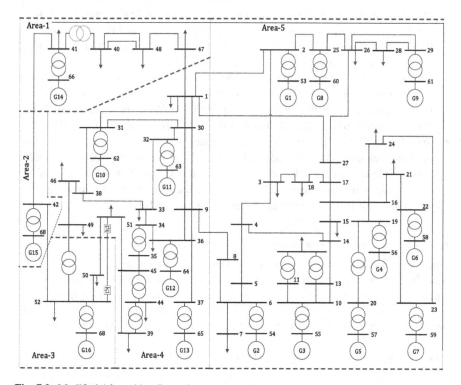

Fig. 7.3 Modified 16-machine 5-area interconnected systems

To improve the interconnection level between Area-3 and -4, an HVDC transmission is installed between Bus-52 (in Area-3) and -51 (in Area-4). The detailed system description including network data and dynamic data of the generators and excitation systems are shown in [6].

7.4.2 Result of Damping Distribution

By modal analysis, the dominant modes are obtained, as shown in Table 7.1. It can be seen that there are 15 dominant modes with weak damping ratios.

The generator participation factors corresponding to these dominant modes can also be obtained by means of modal analysis, as shown in Fig. 7.4. From this figure, we can compare the participation factors of generators to each dominant mode. Furthermore, the dominant participation generators for each mode are obtained as Table 7.2 shows, which provides the reference to configure PSSs and HVDC-WADC to achieve the maximum damping effectiveness on the assigned modes.

The damping distribution of each mode to each PSS or HVDC-WADC can be determined by the following planned procedure:

- *Step-1*: Distribution of the inter-area mode to HVDC-WADC. From Fig. 7.4a, it can be seen that mode-50 is an inter-area mode participated by multimachines. Figure 7.5a shows its mode shape. This mode reflects the oscillations among G13 (in Area-4), G15 (in Area-2), G14 (in Area-1), and G6–G7 (in Area-5), seen in Table 7.2. Since the participation factors of these dominant generators are balanced (see Fig. 7.4a), it is difficult to select a suitable PSS to damp this mode effectively. In this situation, the HVDC-WADC is a relatively better choice, wherefore the task of damping this inter-area mode is assigned to it.
- *Step-2*: Distribution of the typical local modes to PSSs. From Table 7.2, it can be seen that mode-54, -61, -65, -73, -94, and -99 are the typical local modes. Each mode is dominated by only one generator. Figure 7.5b shows one of the mode shapes of these modes. For these modes, it is necessary to assign the corresponding PSSs to damp them, as shown in the Table 7.2.

Table 7.1 Dominant oscillation modes

No. of oscillation mode	Oscillation frequency (Hz)	Damping ratio
Mode-50, -52, -54, -58	0.4446, 0.5803, 0.7028, 0.8240	0.0263, 0.0112, 0.0024, 0.0256
Mode-61, -63, -65, -67	1.2207, 1.2291, 1.2399, 1.3301	−0.0536, −0.0311, −0.0829, 0.0187
Mode-69, -71, -73, -82	1.3469, 1.3709, 1.3832, 1.6067	−0.0344, −0.0553, −0.0194, 0.0115
Mode-90, -94, -99	1.6215, 1.6245, 1.9969	−0.0090, 0.0374, −0.0417

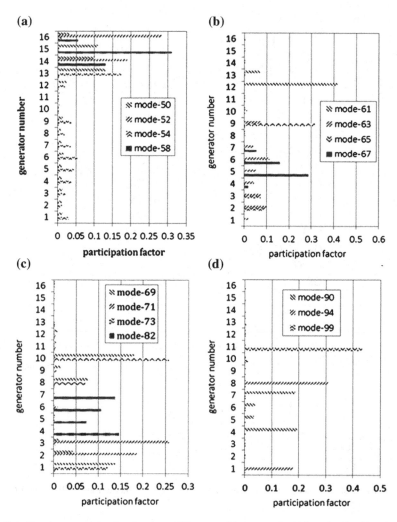

Fig. 7.4 Generator participation factor to different oscillation modes, (**a**) mode-50, -52, -54, -58; (**b**) mode-61, -63, -65, -67; (**c**) mode-69, -71, -73, -82; (**d**) mode-90, -94, -99

- *Step-3*: Distribution of other oscillation modes to other PSSs. As shown in Fig. 7.4 and Table 7.2, the other modes, i.e., mode-52, -63, -67, -69, -71, -82, and -90, indicates the oscillation among multimachines. Unlike mode-50 and -99 mentioned in *Step-1* and *-2*, the generator participation factors on this kind of modes are complex. Figure 7.5c shows a mode shape of this kind of oscillation mode. In general, for this kind of mode, the generator with the highest partici-pation factor on such mode should be configured to the corresponding PSS.

Table 7.2 Assignment of PSSs and HVDC-WADC

No. of LFO mode	Dominant participation generator	Assignment of PSSs and HVDC-WADC
50	G13, G15, G14, G6, G7	HVDC-WADC
52	G16, G14	G16-PSS
54	G13	G13-PSS
58	G14, G16	G14-PSS
61	G12	G12-PSS
63	G6, G2, G9, G3, G5, G7	G2-PSS
65	G9	G9-PSS
67	G5, G6	G5-PSS
69	G10, G1	G1-PSS
71	G3, G2	G3-PSS
73	G10	G10-PSS
82	G4, G7, G6, G5	G4-PSS
90	G4, G7	G7-PSS
94	G8	GPSS
99	G11	G11-PSS

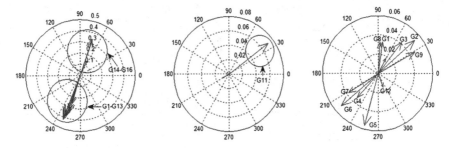

Fig. 7.5 Mode shapes, (**a**) inter-area mode (mode-50); (**b**) local mode (mode-99); (**c**) non-typical LFO mode (mode-63)

However, since sometimes the PSS configured in *Step-2* can also damp this kind of oscillation mode, in this step it need not to configure a PSS for this mode. This can be confirmed by the root-locus analysis. So another PSS can be flexibly selected to damp this kind of mode.

7.4.3 Design Result

The configuration of phase-compensation blocks is finished after 15 steps. In each step, the dominant modes that can be damped by the designed PSS or HVDC-WADC are also pointed out as the reference for other design steps.

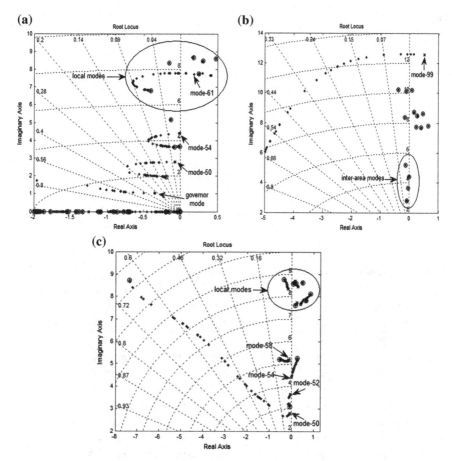

Fig. 7.6 Root-locus with the varying of the control gain of the PSS or HVDC-WADC, (**a**) PSS at G11; (**b**) PSS at G13; (**c**) HVDC-WADC

Figure 7.6 shows the root-locus of the test systems using the designed G11-PSS, G13-PSS, and HVDC-WADC. It can be seen that these PSSs and HVDC-WADC can achieve the effective damping response on their assigned modes as said in Sect. 7.4.2 (Table 7.2).

Furthermore, according to the global optimization proposed in Sect. 7.3.3, the control gains of PSSs and HVDC-WADC can be optimized. Figure 7.7 shows the iteration process of the objective function (7.1). It can be seen from this figure that the iteration is ended when the convergence criteria is satisfied, which means the damping ratios of all the dominant modes are increased to more than 0.05. More specifically, Table 7.3 shows the designed parameters of the PSSs and the HVDC-WADC.

Fig. 7.7 Iterative process

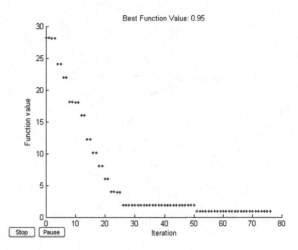

Table 7.3 Parameters of the PSSs and HVDC-WADC

Step	Type	K	Tw	T1	T2	T3	T4	Dominant mode
1	G16-PSS	94	10	0.03	0.02	0.03	0.02	mode-52, -58
2	G13-PSS	93	10	0.05	0.01	0.04	0.01	mode-54, -50, -61
3	G14-PSS	94	10	0.04	0.02	0.04	0.03	mode-52, -58, -50
4	G12-PSS	98	10	0.11	0.01	0.10	0.01	mode-61
5	G2-PSS	92.5	10	0.10	0.02	0.10	0.02	mode-69, -63, -65
6	G9-PSS	98	10	0.06	0.02	0.06	0.02	mode-63, -69, -65
7	G5-PSS	93	10	0.08	0.02	0.08	0.02	mode-67, -82
8	G1-PSS	5	10	0.11	0.02	0.12	0.02	mode-69, -94
9	G3-PSS	93	10	0.09	0.02	0.08	0.02	mode-71
10	G10-PSS	93	10	0.10	0.02	0.09	0.02	mode-73
11	G4-PSS	5	10	0.10	0.02	0.10	0.02	mode-82, -67, -63, -54
12	G7-PSS	20	10	0.09	0.02	0.08	0.02	mode-90, -67, -63, -54
13	GPSS	100	10	0.09	0.01	0.09	0.01	mode-94
14	G11-PSS	93	10	0.08	0.02	0.05	0.02	mode-99
15	HVDC-WADC	3	10	0.05	0.01	0.05	0.01	mode-50

7.4.4   Performance Validation

7.4.4.1   Eigenvalue Analysis

To validate the SDGO method, the eigenvalue analysis is performed on the test system with or without the optimized PSSs and HVDC-WADC. Figure 7.8 shows the result about the dominant eigenvalues. From this figure it is clear that after optimization, all dominant eigenvalues are assigned in the prospective region where

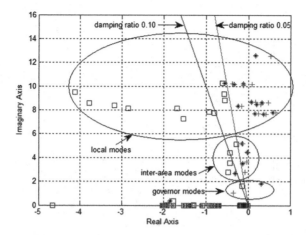

Fig. 7.8 Dominant eigenvalues of the test systems without, with the initial and with the optimized PSSs and HVDC-WADC (In this figure, *plus*, *asterisk*, and *square* is the test systems without, with the initial and with the optimized PSSs and HVDC-WADC, respectively.)

Table 7.4 Dominant eigenvalues of the test systems with or without PSSs and HVDC-WADC

Oscillation mode type	With initial PSSs and HVDC-WADC		With optimized PSSs and HVDC-WADC	
	Dominant eigenvalue	Damping ratio	Dominant eigenvalue	Dominant eigenvalue
Exciter mode	$0.3075 \pm 1.8310i$	-0.1656	$-0.1397 \pm 1.6193i$	0.0860
Inter-area mode	$-0.3089 \pm 2.6691i$	0.1149	$-0.4705 \pm 2.7586i$	0.1681
	$-0.1338 \pm 3.5230i$	0.0380	$-0.4327 \pm 3.5487i$	0.1210
	$-0.0177 \pm 4.4579i$	0.0040	$-0.4314 \pm 4.4132i$	0.0973
	$-0.1447 \pm 5.1745i$	0.0279	$-0.2852 \pm 5.1659i$	0.0551
Local mode	$0.3612 \pm 7.6769i$	-0.0470	$-1.5279 \pm 7.2691i$	0.2057
	$0.1585 \pm 7.7144i$	-0.0205	$-0.8053 \pm 7.7587i$	0.1032
	$0.5756 \pm 7.8006i$	-0.0736	$-0.9243 \pm 7.8664i$	0.1167
	$-0.2393 \pm 7.3248i$	0.0287	$-1.6875 \pm 7.1482i$	0.2028
	$0.1999 \pm 7.4619i$	-0.0236	$-2.8534 \pm 7.1481i$	0.3305
	$0.3218 \pm 7.6487i$	-0.0372	$-0.5591 \pm 7.8129i$	0.0633
	$0.0772 \pm 7.6898i$	-0.0089	$-3.1837 \pm 7.3750i$	0.3553
	$-0.3391 \pm 10.1725i$	0.0333	$-0.5674 \pm 9.3773i$	0.0604
	$-0.1125 \pm 10.2137i$	0.0110	$-3.7882 \pm 7.5973i$	0.4032
	$-0.5583 \pm 10.2545i$	0.0544	$-0.6147 \pm 10.2740i$	0.0597
	$0.1645 \pm 12.5962i$	-0.0131	$-4.1247 \pm 9.4774i$	0.3991

the damping ratio is more than 0.05. Table 7.4 further gives the values of the dominant eigenvalues and corresponding damping ratios. It can be seen that all damping ratios of the dominant modes are increased when the test systems using the optimized PSSs and HVDC-WADC.

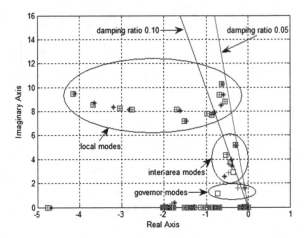

Fig. 7.9 Impacts of operating conditions on the test system with the optimized PSSs and HVDC WADC (In this figure, *square*, *asterisk*, and *plus* is the dominant eigenvalues of the test system on the normal operating condition, on the operating condition-1 and -2, respectively.)

In order to validate the damping performance of the optimized PSSs and HVDC-WADC under different operating conditions, two typical operating conditions are considered, that is, operating condition-1: two tie-lines between Area-4 and -5 are out of service, and operating condition-2: +20 % loading increase on all the buses with P&Q loads. Figure 7.9 shows the results on the dominant eigenvalues of the closed-loop system operating at these conditions. From this figure, it can be seen that no matter the changing of system operating condition, all dominant eigenvalues are always placed in the acceptable region.

7.4.4.2 Nonlinear Simulation

The nonlinear simulation is carried out to further examine the damping performance of the optimized PSSs and HVDC-WADC. In this simulation case, the tie-line 1–2, one of the backbone interconnected lines between Area-4 and -5, shown in Fig. 7.3, is outage at simulation time of 1.0 s to examine the stability of the test systems without, with the initial and with the optimized PSSs and HVDC-WADC, respectively.

Figure 7.10 shows the dynamic responses of the power flow through the tie-lines. It can be seen that for the test systems without or with the initial PSSs and HVDC-WADC, the line outage excites serious power oscillations, which leads to system collapse. But when the test systems are equipped with the optimized PSSs and HVDC-WADC, the oscillations are damped effectively, so the test system can operate stably.

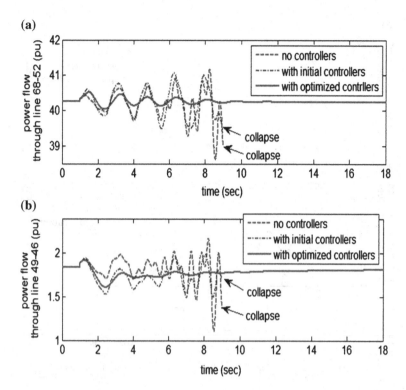

Fig. 7.10 Responses of the power flow, (**a**) through line 652; (**b**) in line 49-46

To investigate the stability at the generation side, especially the oscillation among generators located in different areas, Fig. 7.11 gives the dynamic responses of the relative angle between G7 (in Area-5) and G15 (in Area-2), between G16 (in Area-3) and G14 (in Area-1), and between G10 (in Area-4) and G1 (in Area-5), respectively. From these figures, it is clear that when using the optimized PSSs and HVDC-WADC, the oscillations among generators located in different areas can be damped effectively.

Furthermore, to examine the effectiveness of the HVDC-WADC, Fig. 7.12 gives the dynamic responses of its control-output and the power flow through the HVDC interconnected line. It can be seen from Fig. 7.12a that both the initial and the optimized HVDC-WADC can response to the oscillation damping. However, the optimized HVDC-WADC represents a good control response by the coordination of the optimized LPSSs. From Fig. 7.12b it can be seen that the implementation of HVDC-WADC can increase transmitted capacity of the interconnected line to a certain degree.

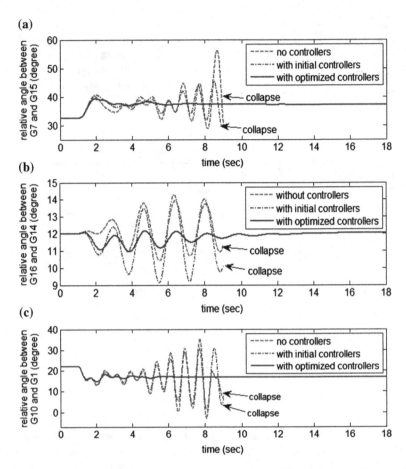

Fig. 7.11 Responses of the relative angle, (**a**) between G7 and G15; (**b**) between G16 and G14; (**c**) between G10 and G1

7.5   Summary

In this chapter, a SDGO method is proposed to optimize local and wide-area controllers simultaneously. Its technological concept and implementation flowchart are described. First, a practical method, which carries out the proper distribution of each dominant damping mode to each PSS and HVDC-WADC, is presented by means of modal analysis. Based on these, the guideline of sequential design is presented to coordinately design the phase-compensation blocks of PSSs and HVDC-WADC step-by-step. The global optimization is proposed to further optimize their control gains. The objective function with the ability of adaptive searching is established, and the pattern search algorithm is used for the nonlinear optimization. Both the eigenvalue analysis and the nonlinear simulation validate

(a)

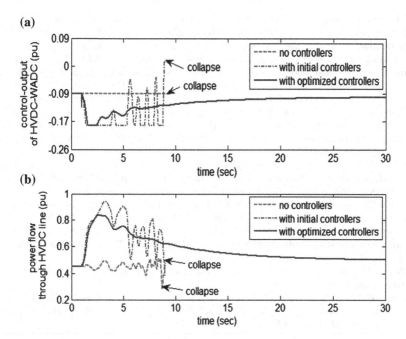

(b)

Fig. 7.12 Responses of the HVDC-WADC and the HVDC line, (**a**) control-output of HVDC-WADC; (**b**) HVDC line

that the designed PSSs and HVDC-WADC can provide effective damping on power oscillations in the AC/DC interconnected systems. The SDGO method can reasonably configure kinds of local and wide-area damping controllers, and achieve overall stability enhancement. It is suitable for stability analysis and controller design for large AC/DC interconnected systems.

References

1. Do Bomfim ALB, Taranto GN, Falcao DM (2000) Simultaneous tuning of power system damping controllers using genetic algorithms. IEEE Trans Power Syst 15(1):163–169
2. Abdel-Magid YL, Abido MA, Al-Baiyat S, Mantawy AH (1999) Simultaneous stabilization of multimachine power systems via genetic algorithms. IEEE Trans Power Syst 14(4):1428–1439
3. Cai LJ, Erlich I (2005) Simultaneous coordinated tuning of PSS and FACTS damping controllers in large power systems. IEEE Trans Power Syst 20(1):294–300
4. Kuiava R, de Oliveira RV, Ramos RA, Bretas NG (2006) Simultaneous coordinated design of PSS and TCSC damping controller for power systems. In: Proceedings of IEEE power engineering society general meeting, Oct 2006. pp 1–4
5. Guo C, Li QZ (2009) Simultaneous coordinated tuning of PSS and FACTS damping controller using improved particle swarm optimization. In: Proceedings of Asia-Pacific power and energy engineering conference, Mar 2009. pp 1–4
6. Rogers G (2000) Power system oscillations. Kluwer, Norwell, MA

7. Kundur P (1994) Power system stability and control. McGraw-Hill, New York
8. Gibbard MJ, Martins N, Sanchez-Gasca JJ, Uchida N, Vittal V, Wang L (2001) Recent applications of linear analysis techniques. IEEE Trans Power Syst 16(1):154–162
9. Kamwa I, Gerin-Lajoie L (2000) State-space system identification-toward MIMO models for modal analysis and optimization of bulk power systems. IEEE Trans Power Syst 15 (1):326–335
10. Arabi S, Rogers GJ, Wong DY, Kundur P, Lauby MG (1991) Small signal stability program analysis of SVC and HVDC in AC power systems. IEEE Trans Power Syst 6(3):1147–1153
11. Korba P, Larsson M, Rehtanz C (2003) Detection of oscillations in power systems using kalman filtering techniques. In: Proceedings of 2003 IEEE conference on control applications, June 2003. pp 183–188
12. Larsen EV, Swann DA (1981) Applying power system stabilizers, part I: general concepts, part II: performance objectives and tuning concepts, part III: practical considerations. IEEE Trans Power Apparatus Syst PAS-100(6): 3017–3046
13. Aboul-Ela ME, Sallam AA, McCalley JD, Fouad AA (1996) Damping controller design for power system oscillations using global signals. IEEE Trans Power Syst 11(2):767–773
14. Kamwa I, Grondin R, Hebert Y (2001) Wide-area measurement based stabilizing control of large power systems-a decentralized/hierarchical approach. IEEE Trans Power Syst 16 (1):136–153
15. Bose A (2010) Smart transmission grid applications and their supporting infrastructure. IEEE Trans Smart Grid 1(1):11–19
16. Jiang ZH, Li FX, Qiao W, Sun HB, Wan H, Wang JH, Xia Y, Xu Z, Zhang P (2009) A vision of smart transmission grids. In: IEEE Proceedings of power engineering society general meeting, July 2009. pp 1–10

Chapter 8
Robust Coordination of HVDC and FACTS Wide-Area Damping Controllers

In this chapter, a wide-area robust coordination method is proposed for HVDC- and FACTS-WADC. It can provide an effective damping on multiple inter-area oscillations excited by various operating conditions. A design procedure is planned as a way of executing the robust coordination for HVDC- and FACTS-WADC. In each step, the WADC design is formulated as a standard robust problem of H_2/H_∞ output feedback control. The proposed coordination method is able to consider the output disturbance rejection problem, reduce the control effort, and ensure robustness against model uncertainties.

8.1 Overview of Wide-Area Damping Control

As a new-type stability control strategy based on WAMS, wide-area damping control can efficiently utilize a controllable device (e.g., PSS, FACTS, and HVDC controllers) to form a wide-area control loop for stability enhancement of interconnected systems. Nowadays, some kinds of control methods, such as the decentralized/hierarchical control [1] and the two-level control [2], have been proposed for designing PSS with local and global control loops. Although PSS-WADC can perform acceptably at some operating condition, in real interconnected systems, there are many generators distributed in different areas, thus the coordination of different PSSs linked to different generators has to be considered carefully. Besides, PSS is usually used to damp local modes, while PSS-WADC mainly deals with inter-area modes. For this reason, the effect of PSS-WADC on local oscillations has to be considered.

WAMS-based FACTS damping control [3–6] is considered to be advantageous for damping inter-area oscillations. However, its damping performance highly depends on the suitable allocation of FACTS devices [7, 8], which is generally oriented toward local control objectives (e.g., reactive power compensation, power flow control, or voltage stability control). Therefore, FACTS-WADC might not provide an acceptable damping for the inter-area oscillation at some operating conditions.

© Springer-Verlag Berlin Heidelberg 2016
Y. Li et al., *Interconnected Power Systems*, Power Systems,
DOI 10.1007/978-3-662-48627-6_8

Unlike PSS and FACTS-WADC, HVDC-WADC is a better choice for damping inter-area oscillation between interconnected areas. Since HVDC transmission is generally used for network interconnection, as a backbone interconnected line, it significantly influences the operating characteristic of interconnected systems. Taking this into consideration, it is feasible to establish WAMS-based HVDC supplementary control to play a role in damping inter-area oscillation in large AC/DC interconnected systems. At present, the HVDC-WADC is being considered in Chinese southern grid (CSG) [9, 10].

However, in practice, there are usually various inter-area oscillation modes that act together on destabilizing interconnected systems. In such a case, a simple FACTS- or HVDC-WADC may not be sufficient to damp multiple inter-area modes. Therefore, the control coordination of multiple WADCs should be considered to establish a wide-area control network (WACN) for the overall stability enhancement.

8.2   Description of Wide-Area Control Networks Using Multiple Power Electronics-Based Controllers

Figure 8.1 shows the architecture of WACN using different kinds of controllable devices (HVDC and FACTS) to implement wide-area coordinated control. The presented five-area system is in fact the NETS–NYPS interconnected system mentioned in Chap. 7. Here, it is modified with an HVDC interconnected line

Fig. 8.1 An architecture of wide-area control network using multiple WADCs

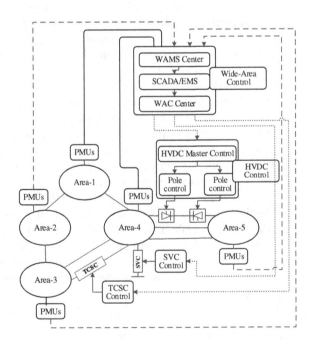

between NETS (Area-5) and NYPS (Area-4), a shunt-type FACTS device (SVC) at one bus of NYPS (Area-4), and a series-type FACTS device (TCSC) between NYPS and the neighboring equivalent area (Area-3).

The wide-area control network mainly includes the following aspects:

- Measurement of wide-area information. Through PMUs placed in optimal locations of each area, various dynamic variables (e.g., bus voltage, line active power flow, speed of remote generator, etc.) are monitored and transmitted from each area to the WAMS center.
- Processing of wide-area information. The useful information is selected and preprocessed in WAMS center, and then sent to SCADA/EMS.
- Implementation of wide-area control strategy. The suitable control signals are selected for HVDC and FACTS controllers, respectively. Then, through the designed centralized controller embedded in WAC center, the control output signals are sent to HVDC and FACTS devices simultaneously.
- Implementation of local control strategy using wide-area information. As the supplementary control linked to the local controller, the wide-area control output can be adopted by the related HVDC and FACTS local controllers to provide stabilizing control for the secure operation of power systems.

8.3 Controller Design Formulation

8.3.1 Multi-objective Synthesis of Wide-Area Robust Control

In order to achieve a robust damping performance on a wide range of operating conditions, the robust design technique, that is the mixed H_2/H_∞ synthesis with regional pole placement, is employed to execute the multi-objective synthesis design of multiple WADCs.

The configuration of multi-objective wide-area robust controller synthesis is shown in Fig. 8.2, where the output channel z_∞ is associated with the H_∞

Fig. 8.2 Configuration of multi-objective robust controller synthesis

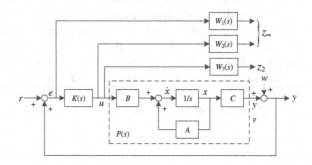

performance while the channel z_2 is associated with the LQG aspects (H_2 performance). Both the low-pass and the high-pass filter, that is the weight functions $W_1(s)$ and $W_2(s)$, are placed in the H_∞ performance channel to ensure the output disturbance rejection and the robustness against model uncertainties, respectively. The high-pass filter or some small constant $W_3(s)$ is placed in the H_2 performance to reduce the control effort.

If the weight function $W_i(s)$ ($i = 1, 2, 3$) goes unconsidered at first, which can later on be placed in a diagonal form using the sdiag function embedded in Matlab [11], the state-space representation of the plant model $P(s)$ can be formulated as:

$$
\begin{bmatrix} \dot{x} \\ \hline z_{1-\infty} \\ \hline z_{2-\infty} \\ \hline z_{3-2} \\ \hline y \end{bmatrix} = \begin{bmatrix} A & 0 & B \\ \hline C & I & 0 \\ 0 & 0 & I \\ \hline 0 & 0 & I \\ \hline C & I & 0 \end{bmatrix} \begin{bmatrix} x \\ w \\ u \end{bmatrix} \tag{8.1}
$$

where A, B, and C are the state, input, and output matrix, respectively; x is the state vector; z is the regulator output; y is the plant output; w is the disturbance input; and u is the plant input.

The main task of the robust design is to find an output feedback controller $K(s)$ that minimizes $\|T_\infty\|_\infty$ subject to $\|T_2\|_2 < h$, where T_∞ and T_2 are the closed-loop transfer functions from w to z_∞ and z_2 respectively. In addition, the closed-loop poles need to be placed in the desired LMI region D. Generally, the design problem is solved by suitably defining the objectives as the argument of the function hinfmix, which is available in the LMI Toolbox of MATLAB [11], with the appropriate pole placement in LMI regions as described in the following section.

8.3.2   Pole Placement in LMI Regions

The transient response of the closed-loop plant is related to the location of poles.

As shown in Fig. 8.3, all closed-loop poles should be limited in the expected region of the left half plane, in order to achieve a good response and at the same time to avoid the fast dynamics and high-frequency gain in the controller. The bound of the desired damping ratio can be defined by setting the inner angle θ and the minimum decay rate $-\sigma$ for the conic sector region. Generally, the function lmireg in MATLAB [11] can be employed for such pole placement in the LMI region. In this way, the function constructs the constraint condition of the pole placement of the closed-loop plant. The result constitutes the solution of the multi-objective control problem by using the mixed H_2/H_∞ synthesis.

Fig. 8.3 LMI region for pole placement

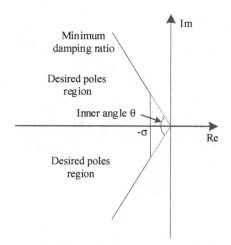

8.4 Design Procedure of Wide-Area Robust Coordinated Control

In order to coordinate multiple HVDC- and FACTS-WADC for achieving optimal robust performance, based on the multi-objective mixed H_2/H_∞ control synthesis mentioned above, a design procedure is proposed. As illustrated in Fig. 8.4, it has the following steps:

- *Step 1*: Linearized system modeling and small-signal stability assessment (SSSA). First of all, the linearized plant model of interconnected systems including multiple HVDC and FACTS devices is established. Here, the model is obtained by a set of differential–algebraic equations containing the dynamic model of HVDC device, FACTS devices, generators, loads, and other equipment and related controllers. A SSSA is carried out to get the system dynamic information including the dominant inter-area modes.
- *Step 2*: Choice and assignment of suitable control signals to each HVDC- and FACTS-WADC. Interconnected systems include numerous operating variables which can be selected as wide-area control signals, such as bus voltage, line-current, and active power flow. According to the SSSA results from the *Step 1*, a suitable selection method is used to choose optimal control inputs. Besides that, during the process of signal selection, the suitable HVDC or FACTS devices capable of damping some dominant inter-area modes are assigned.
- *Step 3*: Sequential design based on the multi-objective mixed H_2/H_∞ controller synthesis. As shown in Fig. 8.4, the first controller (WADC-1) is designed using the plant model obtained in *Step 1* and the optimal control signals obtained in *Step 2*, the applied layout method is the robust synthesis described in Sect. 8.3. Afterwards, based on the WADC-1 and the plant model, the closed-loop system

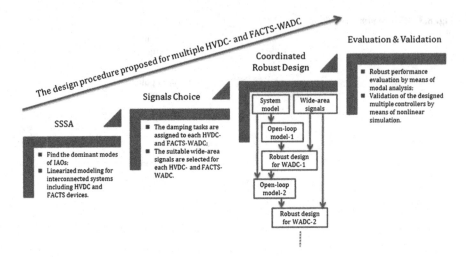

Fig. 8.4 Flowchart of the design procedure

model can be formed, which is used as the open-loop model for the second WADC design. Similarly, once the WADC-2 is obtained, it can be used to form the second closed-loop model for the further controller design. In this way, all HVDC- and FACTS-WADC considered can be designed sequentially.

- *Step 4*: Robust performance evaluation and nonlinear simulation validation. The robustness of the designed HVDC- and FACTS-WADC is evaluated on the full-order closed-loop plant model formed in the last step. Furthermore, the designed WADCs are verified on various operating scenarios such as line-fault, line-outage, and load-shedding.

8.5   Case Study

8.5.1   Choice of Suitable Wide-Area Control Signals

The modal analysis is used to reveal the dynamic behavior of this test system. Figure 8.5 shows the shapes of the dominant oscillation modes. It can be clearly seen that there are four typical inter-area oscillation modes with the low damping ratios: the oscillation between Area-4, -5 and Area-1, -2, -3; between Area-3 and Area-1, -5; between Area-4 and Area-5, and between Area-2 and Area-1, -3. The oscillation frequencies are 0.4278, 0.5734, 0.6723, and 0.8215 Hz, respectively. Table 8.1 gives the detailed information on the oscillations including participation generators, oscillation frequencies f, damping ratios ζ, and so on.

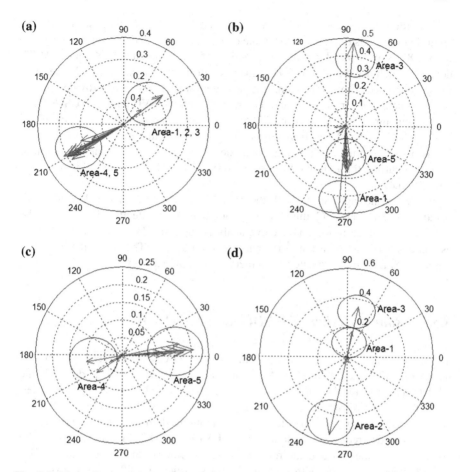

Fig. 8.5 Mode shapes of inter-area oscillation, **a** mode-1, 0.4278 Hz; **b** mode-2, 0.5734 Hz; **c** mode-3, 0.6723 Hz; **d** mode-4, 0.8215 Hz

Table 8.1 Inter-area oscillation modes

Mode index	Eigenvalues	Mode shape	f (Hz)	ζ
1	$-0.1663 \pm 2.6878i$	G14–G16 versus G1–G13 (Area 1–3 vs. Area 4–5)	0.4278	0.0617
2	$-0.0549 \pm 3.6030i$	G1–G9, G14 versus G16 (Area 1, 5 vs. Area-3)	0.5734	0.0152
3	$-0.2939 \pm 4.2245i$	G10–G12 versus G1–9 (Area-4 vs. Area-5)	0.6723	0.0694
4	$-0.1312 \pm 5.1619i$	G15 versus G14, G16 (Area-2 vs. Area 1, 3)	0.8215	0.0254

According to the design procedure proposed in Sect. 8.4, the first task for the controller design is to assign the suitable controller device and choose input signals. Figure 8.6 shows the damping contribution of the optional HVDC and FACTS devices, which use different line-currents as control input signals for each inter-area mode concerned. From Fig. 8.6a, c it can be seen that is able to provide a better damping behavior for modes 1 and 3 compared with the SVC and the TCSC. Thus, the current through line 82 (line 37–65) is selected as the control input of the HVDC-WADC. Similarly, from Fig. 8.6b, d it is clear that the SVC-WADC performs better in terms of damping the modes 2 and 4 by using wide-area signals. Thus, the current through line 85 (line 52–68) is selected as the control input of the SVC-WADC.

It is worth noting that although the TCSC is a candidate for the WACN, compared to HVDC and SVC, this device cannot provide an effective damping for the dominant inter-area modes. The above analysis suggests HVDC and SVC devices as sufficient for damping the dominant inter-area modes. Therefore, these control possibilities are finally selected to implement WADC strategy.

8.5.2   Robust Design of HVDC- and FACTS-WADC

The multi-objective mixed H_2/H_∞ synthesis is employed for the sequential robust design of HVDC- and FACTS-WADC. In order to perform the robust design in the LMI framework, the full-order system should be reduced to an acceptable order. Here, the balanced truncation method is used for the model reduction [2]. For the SVC-WADC design, the open-loop system is reduced from 201th to 11th order. The same order reduction is done for the HVDC-WADC. Figure 8.7 shows the comparison of the full and the reduced-order system. It can be seen that the frequency responses of the both systems do not significantly differ at the given frequency interval.

In order to simultaneously enhance the output disturbance rejection, ensure the robustness against model uncertainties, and reduce the control effort [12–14], suitable weight functions $W_i(s)$ ($i = 1, 2, 3$) should be applied to the output channels z_∞ and z_2, as shown in Fig. 8.2. Here, considering that all dominant inter-area modes are below the frequency 10 rad/s, hence $W_1(s)$ and $W_2(s)$ are determined so that they intersect at around 10 rad/s. Besides, a small constant value is also selected for $W_3(s)$ configured in the H_2 performance channel. The weight functions finally determined for HVDC- and FACTS-WADC are given as follows:

$$W_1(s) = \frac{100}{s + 100}, \; W_2(s) = \frac{100s}{s + 100}, \; W_3(s) = 1 \qquad (8.2)$$

As mentioned before, the HVDC- and FACTS-WADC design can be performed sequentially. However, the order of the designed controller is high, which may be

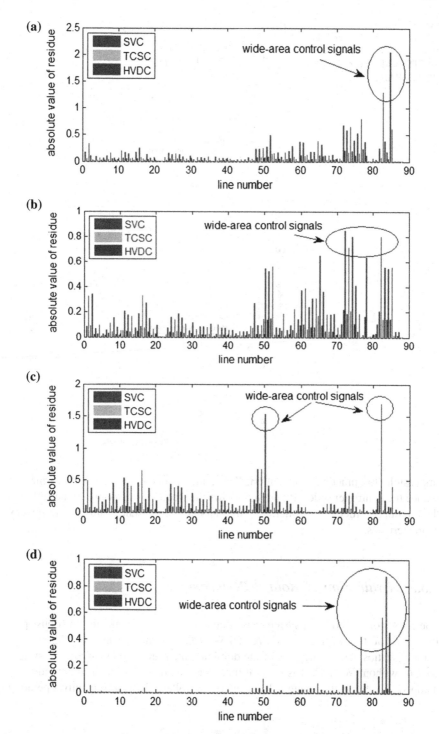

Fig. 8.6 Damping capability of HVDC and FACTS, **a** damping mode-1; **b** damping mode-2;
c damping mode-3; **d** damping mode-4

Fig. 8.7 Frequency responses of the full- and the reduced-order system, **a** for FACTS-WADC; **b** for HVDC-WADC

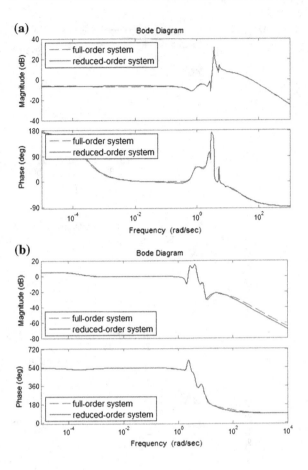

impossible for practical application. Thus, the model reduction is performed to reduce the controller order. Figure 8.8 shows the frequency response of the full- and the reduced-order controller. Both models represent similar frequency characteristics.

8.5.3   Evaluation of Robust Performance

The SSSA of closed-loop system has been carried out to evaluate the robust performance of the HVDC- and FACTS-WADC designed. Table 8.2 shows the damping ratios and frequencies of the dominant inter-area modes of the test system with or without WADC. It is evident that when the dominant inter-area modes are damped effectively by using all designed controller. Considering only operating

Fig. 8.8 Frequency responses of the full and the reduced-order controller, **a** FACTS-WADC; **b** HVDC-WADC

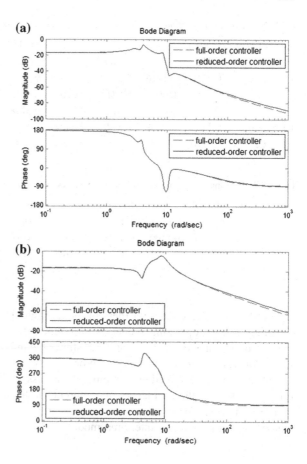

Table 8.2 Dominant inter-area modes of the test system with or without WADC

Mode index	Open-loop		Closed-loop with SVC-WADC		Closed-loop with HVDC-WADC		Closed-loop with SVC and HVDC-WADC	
	f (Hz)	ζ	f(Hz)	ζ	f (Hz)	ζ	f (Hz)	ζ
1	0.4278	0.0617	0.4249	0.0843	0.4234	0.0935	0.4186	0.1080
2	0.5734	0.0152	0.5793	0.0865	0.5730	0.0158	0.5793	0.0867
3	0.6723	0.0694	0.6732	0.0759	0.6796	0.1065	0.6653	0.1145
4	0.8215	0.0254	0.8419	0.0709	0.8216	0.0255	0.8419	0.0716

FACTS-WADC, the damping ratios of mode-2 and -4 are increased greatly, while considering the sole usage of HVDC-WADC, it has a greater impact on the increase of the damping ratios of mode-1 and -3. Such results coincide with the design intention mentioned in Sect. 8.4.

Table 8.3 Dominant inter-area modes of the test system using different-type load

Type of load	Mode-1		Mode-2		Mode-3		Mode-4	
	f (Hz)	ζ	f (Hz)	ζ	f (Hz)	ζ	f (Hz)	ζ
CI	0.4218	0.1216	0.5809	0.0869	0.6605	0.1156	0.8435	0.0784
CP	0.4058	0.0697	0.5714	0.0767	0.6875	0.0805	0.8372	0.0579
CC	0.4170	0.0941	0.5761	0.0855	0.6748	0.1096	0.8403	0.0661
CP + CC	0.4122	0.0808	0.5745	0.0824	0.6844	0.0952	0.8387	0.0616
CP + CI	0.4170	0.0941	0.5761	0.0855	0.6748	0.1096	0.8403	0.0661

Table 8.4 Dominant inter-area modes of the test system with tie-line-outage

Line outage	Mode-1		Mode-2		Mode-3		Mode-4	
	f (Hz)	ζ	f (Hz)	ζ	f (Hz)	ζ	f (Hz)	ζ
1–27	0.4170	0.1100	0.5793	0.0883	0.6494	0.0851	0.8419	0.0714
1–47	0.4186	0.1085	0.5793	0.0867	0.6653	0.1145	0.8419	0.0716
8–9	0.4218	0.1056	0.5777	0.0847	0.7608	0.0682	0.8419	0.0719

In order to evaluate the robustness of the designed HVDC- and FACTS-WADC regarding various operating scenarios, different load types and outages of inter-connected lines are considered. Table 8.3 shows the influence of different load types on the dominant inter-area modes. It can be seen that the designed multiple WADCs can always provide effective damping regardless of the load type. Moreover, Table 8.4 shows the influence of different tie-line-outages on the dominant inter-area modes. These results also represent the effectiveness of the damping behavior toward dominant inter-area modes in the disturbed state.

8.5.4   Nonlinear Simulation

In order to further examine the robustness of the designed HVDC- and FACTS-WADC on damping multiple inter-area oscillation modes, a nonlinear simulation has been carried out on the modified 16-machine 5-area interconnected system.

In this case, the line-to-ground fault near Bus-51 of Line 45–51 (see Fig. 8.3) is simulated. Figure 8.9 shows the dynamic response of the relative angle between different generators located in different areas. Without HVDC- and FACTS-WADC, the occurrence of a line-to-ground fault excites inter-area oscil-lation. Through applying WADC provides an effective oscillation damping.

Furthermore, Fig. 8.10 shows the dynamic responses of the HVDC- and the FACTS-WADC. It can be seen that the designed multiple WADCs can respond for

Fig. 8.9 Responses of the relative angle, **a** between G14 and G16; **b** between G15 and G16

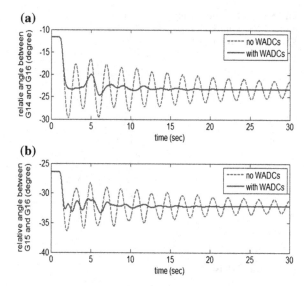

Fig. 8.10 Responses of the controller, **a** output of HVDC-WADC; **b** output susceptance of FACTS-WADC

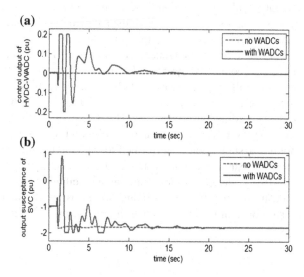

damping power oscillation. The damping process is finished after about 15 s, which indicates the good control performance that the HVDC and FACTS-WADC can achieve. Moreover, Fig. 8.11 shows the dynamic responses of the power flow through different tie-lines. It can also be seen that when implementing multiple WADCs, the power oscillations in the tie-lines can be damped effectively.

Fig. 8.11 Responses of the power flow, **a** through line 52–68; **b** through line 1–47

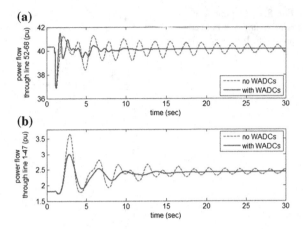

8.6 Summary

This chapter presents a wide-area robust coordination approach for coordinating HVDC- and FACTS-WADC to damp multiple inter-area oscillation modes. This approach can sufficiently utilize the supplementary control functions of HVDC and FACTS devices. Through introducing suitable wide-area signals, the WACN can be constructed with the advanced control ability for enhancing the overall stability of large interconnected systems. The architecture of the WACN is briefly described, and the multi-objective mixed H_2/H_∞ control synthesis is employed to deal with the robust design problem of the HVDC- and FACTS-WADC. A design procedure for robust coordination is proposed. The robustness of the test system installed with the multiple WADCs is evaluated at different operating scenarios. Furthermore, a nonlinear simulation is performed to validate the designed multiple WADCs.

All research results indicate that the wide-area robust coordination approach can effectively damp multiple inter-area oscillation modes and represent a good robustness on different operating scenarios. It is suitable for robust coordinated design of multiple WADCs to enhance the overall stability of large interconnected systems.

References

1. Kamwa I, Grondin R, Hebert Y (2001) Wide-area measurement based stabilizing control of large power systems-a decentralized/hierarchical approach. IEEE Trans Power Syst 16 (1):136–153
2. Dotta D, e Silva AS, Decker IC (2009) Wide-area measurement-based two-level control design considering signals transmission delay. IEEE Trans Power Syst 24(1):208–216

3. Majumder R, Pal BC, Dufour C, Korba P (2006) Design and real-time implementation of robust FACTS controller for damping inter-area oscillation. IEEE Trans Power Syst 21 (2):809–816
4. Majumder R, Chaudhuri B, Pal BC (2005) A probabilistic approach to model-based adaptive control for damping of interarea oscillations. IEEE Trans Power Syst 20(1):367–374
5. Chaudhuri B, Pal BC (2004) Robust damping of multiple swing modes employing global stabilizing signals with a TCSC. IEEE Trans Power Syst 19(1):499–506
6. Chaudhuri NR, Ray S, Majumder R, Chaudhuri B (2010) A new approach to continuous latency compensation with adaptive phasor power oscillation damping controller (POD). IEEE Trans Power Syst 25(2):939–946
7. Farsangi MM, Song YH, Lee KY (2004) Choice of FACTS device control inputs for damping inter-area oscillations. IEEE Trans Power Syst 19(2):1135–1143
8. Farsangi MM, Nezamabadi-pour H, Song YH, Lee KY (2007) Placement of SVCs and selection of stabilizing signals in power systems. IEEE Trans Power Syst 22(3):1061–1071
9. Mao XM, Zhang Y, Guan L, Wu XC (2006) Coordinated control of inter-area oscillation in the China Southern power grid. IEEE Trans Power Syst 21(2):845–852
10. Mao XM, Zhang Y, Guan L, Wu XC, Zhang M (2008) Improving power system dynamic performance using wide-area high-voltage direct current damping control. IET Gener Trans Distrib 2(2):245–251
11. Gahinet P, Nemirovski A, Laub A, Chilali M (1995) LMI control toolbox for use with Matlab: The Math Works Inc., Natick
12. Rogers G (2000) Power system oscillations. Kluwer, Norwell
13. Zhang XP, Rehtanz C, Pal B (2006) Flexible AC transmission systems: modeling and control. Springer, Berlin
14. Zhang Y, Bose A (2008) Design of wide-area damping controllers for inter-area oscillations. IEEE Trans Power Syst 23(3):1136–1143

Chapter 9
Assessment and Choice of Input Signals for Multiple Wide-Area Damping Controllers

In this chapter, a hybrid method is proposed to assess and select suitable input signals for multiple WADCs. In this method, the residue analysis is used to pres-elect a set of input signal candidates from many wide-area/local signals, and the relative gain array (RGA) method is used to finally determine the input signals for multiple WADCs. The advantage of the hybrid method proposed is to save the time of signal selection and at the same time to reduce even to avoid controller inter-action. So if the input signals are selected properly for multiple WADCs, each WADC can be designed independently without considering controller coordination.

9.1 Overview of Signal Selection Methods

The application of WAMS makes it possible to select an optimal control-input from many remote/local operating variables to achieve improved observability and controllability for wide-area stability control. Therefore, a suitable signal selection method should be considered for WADC. For multiple WADCs it can occur that due to interaction among multiple WADCs the acceptable performance cannot be reached even if each WADC has selected the optimal control-input. For this reason, the issue of how to choose multiple input signals should be considered for multiple WADCs in order to reduce or eliminate controller interaction.

Up to now, several methods have been proposed for the selection of control signals. The most common one is the residue method [1–3]. This method is effective to select control signal for simple controller, but cannot evaluate different types of signals (e.g., line flow, bus-voltage angle, generator rotor speed, etc.) for multiple controllers. To overcome this problem, the geometric method [4] has been used to select the control signal for SVC and synchronous condenser (SC). Nevertheless, this method cannot consider the controller interaction sufficiently. In [5, 6], the minimum singular values (MSV), the right-half plane zeros (RHP-zeros), the RGA, and the Hankel singular values (HSV) were employed as indicators to

© Springer-Verlag Berlin Heidelberg 2016
Y. Li et al., *Interconnected Power Systems*, Power Systems,
DOI 10.1007/978-3-662-48627-6_9

find the control-inputs and the allocation of FACTS devices. In summary, the present methods are mainly used to find the suitable control signals for the local damping control strategy, but for the WADC strategy, there are few works reported.

9.2   Description of Relative Gain Array and Residue Analysis

9.2.1   Power System Model

Figure 9.1 shows the basic structure of a power system with multiple WADCs. In this study, the power system is a typical multi-areas interconnected system, where an HVDC link and a FACTS device are installed to improve transmission capacity and voltage stability. The details have been presented in Sect. 7.4.1. The multiple WADCs select system operating variables as the control-inputs. Their control-outputs are the supplementary control-inputs of HVDC and FACTS local controller.

The HVDC link adopts a constant direct-current control mode at the rectifier side, and the FACTS controller adopts a constant bus-voltage control mode. The plant model of this power system is obtained by linearizing system at the normal operating point, and it has a standard state-space form.

$$\begin{cases} \dot{x}(t) = Ax(t) + Bu(t) \\ y(t) = Cx(t) \end{cases} \tag{9.1}$$

where $x(t) \in R^n$ is the state vector; $y(t) \in R^q$ is the output vector related to wide-area measurement signals and $u(t) \in R^m$ is the input vector linked to the control-output signals of multiple WADCs. $A \in R^{n \times n}$, $B \in R^{n \times m}$ and $C \in R^{q \times n}$ are state, input, and output matrices, respectively.

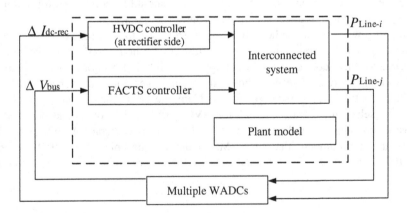

Fig. 9.1 Power system with multiple WADCs

9.2.2 Residue Analysis Method

The residue matrix of the plant model can be obtained from the transfer function representation of (9.1), that is,

$$G(s) = \frac{Y(s)}{U(s)} = C(sI - A)^{-1}B = \sum_{i=1}^{n} \frac{R_i}{(s - \lambda_i)} \tag{9.2}$$

where R_i is the residue of $G(s)$ at eigenvalue λ_i associated with mode i, which can be further expressed as:

$$R_i = C\Phi_i\Psi_iB \tag{9.3}$$

where Φ_i and Ψ_i are the right and the left eigenvector associated with the eigenvalue λ_i, respectively. Both of them are also the i-th element of modal matrix Φ and Ψ related with state matrix A defined in (9.1).

By means of residue analysis, the sensitivity of mode i (eigenvalue λ_i) to the feedback of the open-loop system $G(s)$ from its output to its input can be evaluated conveniently. Therefore, the residue method is useful to find an effective feedback signal that represents high impact on the considered oscillation mode.

9.2.3 RGA Analysis Method

For a multi-input multi-output (MIMO) system, the RGA is a very useful tool to analyze the interaction among different controllers [7]. It can find suitable input–output pairs that significantly reduce or even eliminate interaction among multiple controllers.

9.2.3.1 Definition of RGA

Figure 9.2 shows a typical multi-variable closed-loop control system, where system $G(s)$ has n inputs and m outputs. The relative gain λ_{ij} between input u_j and output y_i is defined as

$$\lambda_{ij} = \frac{(\partial y_i / \partial u_j)|_{\Delta u_k = 0, k \neq j}}{(\partial y_i / \partial u_j)|_{\Delta y_k = 0, k \neq i}} \tag{9.4}$$

where, $(\Delta u_k = 0, k \neq j)$ expresses all uncontrolled outputs, and $(\Delta y_k = 0, k \neq i)$ expresses that the control pair $u_j \rightarrow y_i$ is controlled and at the same time the other control pairs have an ideal control performance.

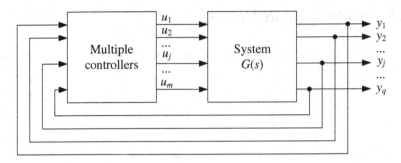

Fig. 9.2 Multi-variable closed-loop control system

It can be seen from (9.4) that λ_{ij} is in fact the ratio of the uncontrolled to the controlled gain. We can use λ_{ij} to evaluate the impact of the other control pairs on the control pair $u_j \to y_i$.

9.2.3.2   Calculation of RGA

Considering a multi-variable system, all the relative gains with the similar representation like (9.4) can construct a whole matrix, which can be obtained by

$$\Lambda_{\text{RGA}} = \left[\lambda_{ij}\right] = G(s) \otimes \left(G(s)^{-1}\right)^{T} \tag{9.5}$$

where $\otimes$ is the multiplication of Hadamard.

It is noted that when $s = 0$, (9.5) calculates the steady-state value of λ_{ij}, and when $s = j\omega = j2\pi f$, (9.5) calculates the dynamic value of λ_{ij} at the specific frequency f. This paper will use the steady-state calculation to select several schemes about the effective input signals of multiple WADCs, and further use the frequency domain analysis to determine the optimal input signals.

More specifically, considering a 2-input 2-output system $G(s)$, its transfer function is represented as

$$y_1(s) = g_{11}(s)u_1(s) + g_{12}(s)u_2(s) \tag{9.6}$$

$$y_2(s) = g_{21}(s)u_1(s) + g_{22}(s)u_2(s) \tag{9.7}$$

where $g_{ij}(s)$ is the element of the system $G(s)$.

When $\Delta y_2 = 0$, according to (9.6) and (9.7), the $y_1(s)$ can be expressed as follows:

$$y_1(s) = \frac{g_{11}(s)g_{22}(s) - g_{12}(s)g_{21}(s)}{g_{22}(s)} \Delta u_1(s) \tag{9.8}$$

According to the definition of RGA, the relative gain λ_{11}, which evaluates the effect of other pairs to control pair $u_1 \rightarrow y_1$, can be expressed as follows:

$$\lambda_{11} = \frac{g_{11}(s)}{(y_1(s)/u_1(s))|_{y_2(s)=0}} = \frac{g_{11}(s)g_{22}(s)}{g_{11}(s)g_{22}(s) - g_{12}(s)g_{21}(s)} \quad (9.9)$$

Similarly, the evaluation on the control pair $u_1 \rightarrow y_2$, $u_2 \rightarrow y_1$ and $u_2 \rightarrow y_2$ can also be calculated as follows:

$$\begin{cases} \lambda_{22} = \lambda_{11} \\ \lambda_{12} = \lambda_{21} = 1 - \lambda_{11} \end{cases} \quad (9.10)$$

9.2.3.3   Properties of RGA

Unlike those purposes [5, 8] of using RGA to find allocation and local control signals for several PSS and/or FACTS devices, in this chapter, the RGA is used to choose optimal input signals for multiple WADCs. The properties of RGA that can be utilized for this purpose are summarized as follows:

- Considering a control pair $u_j \rightarrow y_i$, when its relative gain λ_{ij} is equal to 1 ($\lambda_{ij} = 1$), there will be no effect of other control pairs on this one, which means there are no coupling effects among different control-loops formed by different controllers.
- When λ_{ij} is less than 0 ($\lambda_{ij} \leq 0$), there are serious interactions among different control-loops, and the MIMO control system is unstable.
- When λ_{ij} is between 0.8 and 1.2 ($0.8 \leq \lambda_{ij} \leq 1.2$), there are only small interactive effects among different control-loops. If the suitable design method is implemented for each controller, it can realize the separate design that does not need to consider the coupling effects among different controllers;

When λ_{ij} is less than 0.8 or more than 1.2 ($\lambda_{ij} < 0.8$ or $\lambda_{ij} > 1.2$), there are relatively higher interactive effects among different control-loops. The coordination design is needed to realize the decoupling among different controllers.

9.3   Signal Selection Procedure

For multiple WADCs, once their input signals cannot be selected properly, the interactive effects among different control-loops cannot be avoided. On the other hand, the unsuitable selection may also bring the design difficulty of multiple controllers. To overcome these problems, an implementation procedure based on the residue analysis and the RGA approach is planned in this section to choose suitable input signals for multiple WADCs. This procedure is also available for the signal selection of other multiple controllers.

Figure 9.3 shows the basic flowchart of the hybrid method proposed. More specifically, it mainly includes the following steps:

- *Step 1. Stability assessment of large-scale interconnected systems installed with HVDC and FACTS devices.* The modal analysis is used to reveal the dominant oscillation modes endangering the stability of interconnected systems. The linearized system model is also obtained for the next-step pre selection of input signals.
- *Step 2. Assign damping task to each WADC, and preselect candidates of input signals for each WADC.* In practice, there are more than one dominant LFO modes to affect the stability. Here, the residue analysis is employed to evaluate the damping ability of each WADC on these dominant modes. If one WADC represents high impact on some dominant modes, it is assigned to damp those modes. Further, the relative residue ratio (*RRR*) is proposed in this step. For each WADC, a signal is preselected as one of the candidates for this WADC if it has a high *RRR* value. In this way, a set of signal candidates is selected for multiple WADCs.

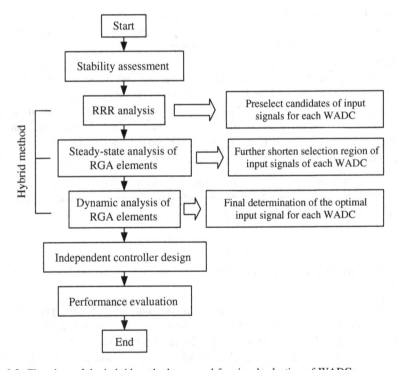

Fig. 9.3 Flowchart of the hybrid method proposed for signal selection of WADCs

- *Step 3. Calculation of RGA and selection of optimal control pairs.* The steady-state values of RGA are calculated to evaluate the interactive effects among different control-loops. Further, the dynamic characteristic of RGA is analyzed by frequency domain method to finally determine the optimal input signals for multiple WADCs.
- *Step 4. Controller designs of multiple WADCs.* The multiple WADCs adopt the suitable control signals determined by Step 2 and 3. The value of λ_{ij} can be used to judge whether the multiple WADCs are designed independently or not. If $0.8 \le \lambda_{ij} \le 1.2$, then each WADC can be designed independently. Otherwise, the controller coordination should be considered carefully.
- *Step 5. Evaluation of damping performance by linear analysis.* The damping performance of the designed multiple WADCs should be evaluated by modal analysis. If the closed-loop system still has an unstable oscillation mode, the controller parameters obtained in Step 4 should be adjusted.
- *Step 6. Validation of damping performance by using nonlinear simulation.* The damping performance should be validated considering different operating conditions and different input signals. Here, the nonlinear simulation is used to validate the damping performance of multiple WADCs.

9.4 Case Study

To illustrate the proposed hybrid method about choosing suitable input signals for multiple WADCs, a detailed case study is carried out based on the 16-machine 5-area interconnected system shown in Fig. 9.3. To improve the interconnected ability, an HVDC link is installed between bus-1 (in area-4) and -2 (in area-5), and a shunt-type FACTS device (SVC) is installed at bus-51 (in area-4). Both HVDC and FACTS devices are equipped with the supplementary WADC for stability enhancement of the interconnected systems.

The modal analysis has been performed to assess system stability. As Table 9.1 shows, there are two dominant modes with small damping ratios ($\zeta < 0.05$). The multiple WADCs are used to damp these inter-area oscillations. To facilitate the description, the denotation mode 1 and 2 is used in the following to, respectively, replace mode 2 and 4 defined in Table 9.1.

Table 9.1 Impact of wide-area signal candidates to the inter-area modes

	HVDC-WADC	FACTS-WADC
Candidate line	Line-85, -83, -77, -72, -74	Line-84, -83, -85, -77, -76
RRR	0.1208, 0.0763, 0.0470, 0.0401, 0.0380	0.2324, 0.1518, 0.1214, 0.1127, 0.0453

9.4.1 Preselection of Input Signal Candidates

According to the selection procedure proposed in Sect. 9.3, the residue analysis method is used to preselect input signal candidates from large numbers of wide-area and local operating variables. These candidates are the control-inputs of multiple WADCs.

For the whole lines of the studied system shown in Fig. 7.3, correspondingly there are 88 line power flows that can be selected as the input signal of each WADC. Totally, there are 88 × 88 = 7744 schemes for the multiple WADCs to select the input signals. In order to preselect a set of effective input signal candidates from the large numbers of schemes, the RRR is proposed in this chapter. More specifically, for the power flow through the i-th line, its RRR_i is defined as

$$\mathrm{RRR}_i = \frac{|R_i|}{\sum_{i=1}^{N} |R_i|} \tag{9.11}$$

where R_i is the residue of power flow through the i-th line; and N is the number of the lines in power systems, here, the studied system has 88 transmission lines, thus $N = 88$.

The RRR value is adequate to evaluate the contribution of each input signal to damp the considered oscillation mode. Figure 9.4 shows the RRR when each line

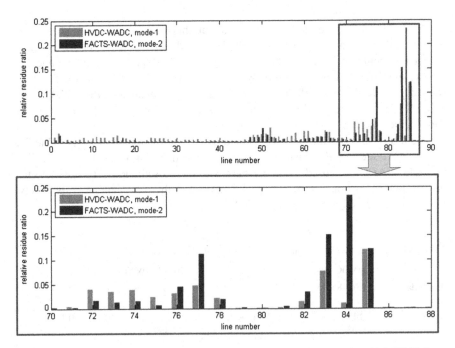

Fig. 9.4 Relative residue ratio of different line power as the input signal of multiple WADCs

Table 9.2 Impact of local signal candidates to the inter-area oscillations

	HVDC-WADC	FACTS-WADC
Candidate line	Line-4 and -5	Line-74 and -72
RRR	0.0052, 0.0035	0.0153, 0.0163

power flow is selected as the input signal of HVDC- and FACTS-WADC. From the enlargement part of this figure, it can be found that for the HVDC-WADC, the power flows in line-85, -83, -77, -72, and -74 have the relatively high RRR values, and for the FACTS-WADC, the power flows in line-84, -83, -85, -77, and -76 have the relatively high RRR values. Therefore, the input signals of multiple WADCs can be selected from these line power flows. Moreover, Table 9.1 gives the detailed *RRR* value of each candidate input signal.

In addition, Table 9.2 shows the impact of local signal candidates to the dominant inter-area oscillations. By comparing with Table 9.1, it is clear that wide-area signals have higher RRR values than local signals, which indicates the better damping ability of WADC than that of local damping control.

9.4.2 Final Choice of Effective Input Signals

According to the selection procedure proposed in Sect. 9.3, the RGA analysis is used to further select the effective input signals for multiple WADCs. The RGA calculation is based on the preselected input signal candidates shown in Table 9.1.

Table 9.3 shows the calculation results about the relative gain when the multiple WADCs adopt different input signals. From this it can be seen that when HVDC- and FACTS-WADC, respectively, adopt $P_{\text{Line-77}}$ and $P_{\text{Line-84}}$ as the control-input, the relative gains are $\lambda_{11} = \lambda_{22} = 1.1978$. Similarly, when they adopt $P_{\text{Line-77}}$ and $P_{\text{Line-85}}$, then $\lambda_{11} = \lambda_{22} = 0.9562$. According to the properties of RGA mentioned in subsection 0, there will be only small controller interactive effects if HVDC- and FACTS-WADCs adopt one of these two control pairs. Thus, the effective input signals can be considered from these two control pairs.

Table 9.3 Candidate lines with high influence to dominant inter-area oscillations

Set no.	HVDC-WADC	FACTS-WADC	Relative Gain ($\lambda_{11} = \lambda_{22}$)
1	Line-85, -85, -85, -85	Line-84, -83, -77, -76	0.2172, 0.1237, 0.0438, 0.0948
2	Line-83, -83, -83, -83	Line-84, -85, -77, -76	2.0361, 0.8763, −0.4806, −2.8719
3	Line-77, -77, -77, -77	Line-84, -83, -85, -77	1.1978, 1.4806, 0.9562, 1.7781
4	Line-72,-72,-72,-72, -72	Line-84, -83, -85, -77, -76	−0.8584, −0.3073, 0.6248, −0.0826, −0.2111
5	Line-74, -74, -74, -74, -74	Line-84, -83, -85, -77, -76	−3.9938, −0.6862, 0.7425, −0.1522, −0.4324

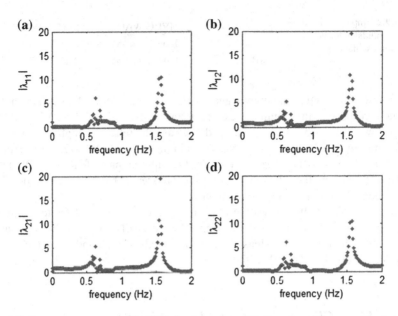

Fig. 9.5 Frequency response of the RGA for the system with $P_{\text{Line-77}}$ and $P_{\text{Line-84}}$ as the input of multiple WADCs, respectively, **a** λ_{11}; **b** λ_{12}; **c** λ_{21}; **d** λ_{22}

Further, by analyzing the dynamic behavior of the RGA elements, the final choice of the input signals can be obtained from the two aforesaid control pairs. Specifically, Figs. 9.5 and 9.6 give the frequency responses of the RGA elements for the system with one of these two control pairs. By comparison, it can be seen that:

- For these two control pairs, all the RGA elements are sensitive to the dominant oscillation modes. It indicates that if the HVDC- and FACTS-WADC select one of the control pairs, the inter-area oscillations can be damped effectively. Moreover, it is clear that at the frequencies of 0.5734 and 0.8215 Hz, the absolute values of the RGA shown in Fig. 9.6 are higher than those shown in Fig. 9.5, which shows that the control pair $u_{\text{HVDC-WADC}} \rightarrow P_{\text{Line-77}}$ and $u_{\text{FACTS-WADC}} \rightarrow P_{\text{Line-85}}$ can achieve more effective damping than the control pair $u_{\text{HVDC-WADC}} \rightarrow P_{\text{Line-77}}$ and $u_{\text{FACTS-WADC}} \rightarrow P_{\text{Line-84}}$.
- At the local oscillation frequency of around 1.5 Hz, the RGA elements in Fig. 9.5 have higher values than those in Fig. 9.6. This indicates that comparing with the control pair $u_{\text{HVDC-WADC}} \rightarrow P_{\text{Line-77}}$ and $u_{\text{FACTS-WADC}} \rightarrow P_{\text{Line-85}}$, the control pair $u_{\text{HVDC-WADC}} \rightarrow P_{\text{Line-77}}$ and $u_{\text{FACTS-WADC}} \rightarrow P_{\text{Line-84}}$ is more sensitive to this local mode. However, it should be noted that the multiple HVDC- and FACTS-WADC are mainly used to damp the inter-area modes not the local mode, thus if they adopt $P_{\text{Line-77}}$ and $P_{\text{Line-84}}$ as the control-input signals, the disturbance of the multiple WADCs to the local mode cannot be avoided. But for the control pair $u_{\text{HVDC-WADC}} \rightarrow P_{\text{Line-77}}$ and $u_{\text{FACTS-WADC}} \rightarrow P_{\text{Line-85}}$, there is no problem like this.

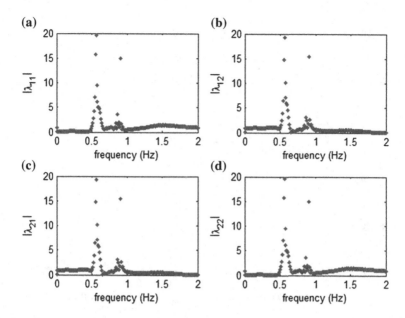

Fig. 9.6 Frequency response of the RGA for the system with $P_{\text{Line-77}}$ and $P_{\text{Line-85}}$ as the input of multiple WADCs, respectively, **a** λ_{11}; **b** λ_{12}; **c** λ_{21}; **d** λ_{22}

According to the preceding analysis, we can see that if the HVDC- and FACTS-WADC adopt $P_{\text{Line-77}}$ and $P_{\text{Line-85}}$ as their control-input signals, they can effectively damp the dominant inter-area oscillations and at the same time avoid the effects on the local mode. Thus, $P_{\text{Line-77}}$ and $P_{\text{Line-85}}$ are finally determined as the control-input signals of HVDC- and FACTS-WADC, respectively.

9.4.3 Comparison with Local Control and Other Wide-Area Control Pairs

To verify the effectiveness of the choice of the input signals, by means of the dynamic analysis of RGA elements, the local control and other wide-area control pairs are compared with the wide-area control signals selected in the above subsection.

For the HVDC- and FACTS-WADC, the power flows in line 2–25 and in line 50–51, which are shown as $P_{\text{Line-4}}$ and $P_{\text{Line-74}}$ in Fig. 7.3, are typical local control signals for HVDC and FACTS-WADC, respectively. For these local signals, Fig. 9.7 shows the frequency response of the RGA elements. Compared with Fig. 9.6, the HVDC- and FACTS-WADC using local control strategy are less sensitive to the dominant inter-area oscillation modes and at the same time they are more sensitive to

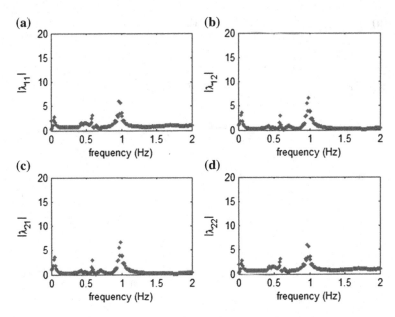

Fig. 9.7 Frequency response of the RGA for the system with $P_{\text{Line-4}}$ and $P_{\text{Line-74}}$ as the input of multiple WADCs, respectively, **a** λ_{11}; **b** λ_{12}; **c** λ_{21}; **d** λ_{22}

the other local mode around 1.0 Hz. It indicates that the local control strategy cannot provide more effective damping for the inter-area modes even affect other local mode.

In addition, Fig. 9.8 gives the frequency response of the RGA elements for the system with $P_{\text{Line-83}}$ and $P_{\text{Line-84}}$ (see Table 9.2) as the input signals of HVDC- and FACTS-WADC. It is shown that if the multiple WADCs adopt other wide-area signals and not the optimal wide-area signals determined in the above subsection, their damping abilities are not sensitive to the considered inter-area modes (0.5734 and 0.8215 Hz, see Table 9.1), and at the same time cause other inter-area oscillations. This further verifies the effectiveness of the hybrid method proposed for signal selection.

9.4.4 Design of Multiple HVDC- and FACTS-WADCs

There are various design methods [9–11] for power system damping controller. Here, the phase compensation method based on residue analysis is used to design multiple WADCs. As discussed in the above subsection, the controller interaction can be reduced significantly when the WADCs adopt optimal input signals, thus a coordinated design for multiple WADCs is not needed. For this reason, each HVDC- and FACTS-WADC is designed separately. After that, the linear analysis is

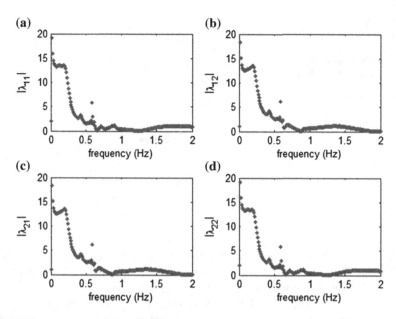

Fig. 9.8 Frequency response of the RGA for the system with $P_{\text{Line-83}}$ and $P_{\text{Line-84}}$ as the input of multiple WADCs, respectively, **a** λ_{11}; **b** λ_{12}; **c** λ_{21}; **d** λ_{22}

performed to assess the damping performance of the designed WADCs. If performance is not acceptable, the coordinated design will be considered. Basically, by phase compensation, the damping controller can be determined with the following transfer function representation:

$$H(s) = K \frac{1}{1+sT_m} \frac{sT_w}{1+sT_w} \left[\frac{1+sT_{\text{lead}}}{1+sT_{\text{lag}}}\right]^{\text{mc}} \tag{9.12}$$

where K is the constant gain; T_m is a measurement time constant; T_w is the washout time constant; T_{lead} and T_{lag} are the lead and the lag time constant and mc is the number of the compensation blocks.

By residue analysis mentioned in Sect. 9.2.2, the residue phase of each WADC can be obtained. The compensation loop like (9.12) is used to partly or to totally compensate this phase. Besides, the root locus analysis is used to determine the control gain K. The designed controller parameters of multiple WADCs are shown in Table 9.4.

Table 9.4 Parameters of HVDC- and FACTS-WADCs

Controller type	K	T_m	T_w	T_1	T_2	mc
HVDC-WADC	0.071	0	10	0.5	0.1	3
FACTS-WADC	0.110	0.07	10	0.5	0.1	1

Fig. 9.9 Dominant
eigenvalues of the studied
interconnected system
(*plus* and *open diamond*
express the systems without
or with HVDC- and
FACTS-WADCs,
respectively)

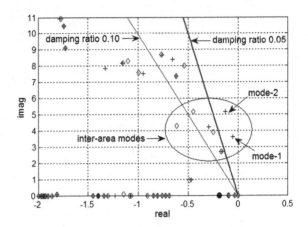

Figure 9.9 shows the dominant eigenvalues of the studied system with or without
HVDC- and FACTS-WADC. As discussed in Sect. 9.4.1, there are two inter-area
modes with weak damping ratios. However, from this figure it is clear that with the
implementation of HVDC- and FACTS-WADC, these modes are placed in the
acceptable region of the *S*-plane. Here, it should be noted that this acceptable
damping performance is obtained without any coordinated design for multiple
WADCs, which indicates that the selected input signals reduce the interaction
among multiple controllers.

9.4.5   Validation of Control Performance

The following cases are investigated to demonstrate the damping performance of
multiple WADCs using the optimal wide-area signals as their control-inputs.

9.4.5.1   Case 1: Damping Performance of Multiple WADCs

In this case, three conditions are considered, that is to say, condition 1: system without
multiple WADCs; condition 2: system with multiple WADCs (no coordinated design,
the controller parameters are shown in Table 9.4); and condition 3: system with
multiple WADCs obtained by a sequential coordinated design mentioned in Chap. 8.
The wide-area signals P_{77} and P_{85}, finally determined in Sect. 9.4.2, are used as the
control-input of HVDC- and FACTS-WADC, respectively.

A big disturbance (outage of line 1–47) is simulated to test the stability of the
system with multiple WADCs. Figure 9.10 gives the system dynamic responses. It
can be seen that the multiple WADCs can damp power oscillations effectively.
Compared to the WADCs determined by the coordinated design method, the
WADCs using the independent design method can also reach the similar damping

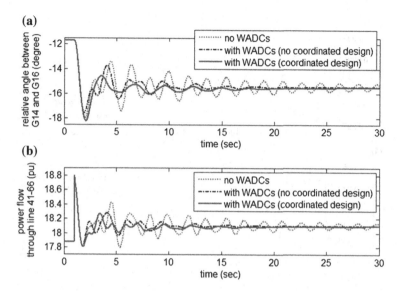

Fig. 9.10 Responses of the test system with multiple WADCs, **a** relative angle between G14 and G16; **b** power flow through line 41–66

performance. Because the selection of suitable input signals reduces the interaction of multiple control-loops, it is successful to design each WADC separately. Since the interaction of multiple WADCs has been considered when selecting the optimal wide-area signals, it does not need the robust design method for controller coordination but directly use the classic phase compensation method to design multiple WADCs, which reduces the difficulty of multiple WADCs design.

9.4.5.2 Case 2: Influence of Wide-Area Control to the Local Control

In order to illustrate the influence of wide-area damping control to the local operating variables related to HVDC and FACTS devices, in this case, the voltage profiles of the HVDC and the FACTS sides are illustrated via comparing the test system with and without multiple WADCs obtained by the independent design method. Figure 9.11 shows the dynamic responses of the voltage on bus-2 (HVDC bus) and on bus-51 (FACTS bus). It is clear that the line outage excites voltage swings at the HVDC and FACTS sides, but the implementation of multiple WADCs prevent the swings successfully and improve the voltage profiles of the HVDC- and the FACTS-bus.

9.4.5.3 Case 3: Robustness at Different Operating Scenarios

Besides the line outage discussed in Case 1 and 2, in this case, another two operating scenarios are considered to verify the robustness of the designed WADCs

(a)

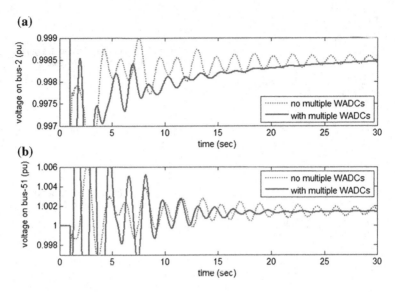

Fig. 9.11 Responses of the test system with multiple WADCs; **a** voltage at bus-2; **b** voltage at bus-57

using the optimal input signals. That is to say, scenario 1: load shedding at bus-46; and scenario 2: line-to-ground fault near bus-8 of line 8–9.

Figures 9.12 and 9.13 show the dynamic responses of the test system at these two operating scenarios. It is clear that in both changes of operating conditions, the

(a)

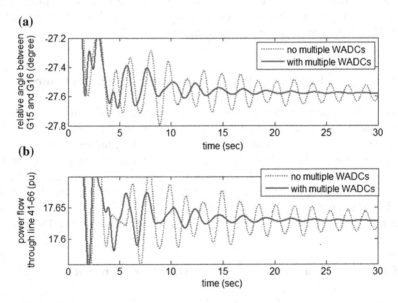

Fig. 9.12 Responses of the test system at operating scenario 1, **a** relative angle between G15 and G16; **b** power flow through line 41–66

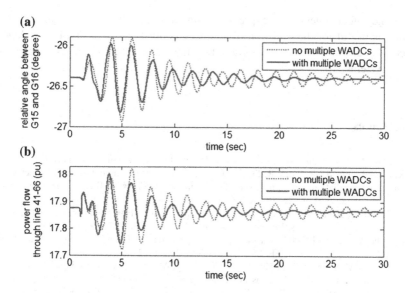

Fig. 9.13 Responses of the test system at operating scenario 2, **a** relative angle between G15 and G16; **b** power flow through line 41–66

system with multiple WADCs always maintains stability, which validates the robustness of the WADCs using the optimal input signals.

9.5   Summary

In this chapter, a hybrid method about selecting suitable input signals for multiple WADCs is proposed. First, the RRR calculation is proposed to preselect input signal candidates from large numbers of wide-area and local signals, and then the RGA calculation, including steady-state value and frequency response, is used to determine the optimal control pairs that can greatly reduce or even can eliminate the interaction among multiple controllers. The proposed method can save time for signal selection and can reduce difficulties of controller design with respect to coordination. Therefore, the proposed method is useful for the stability analysis and controller design of large-scale interconnected systems including abundant wide-area and local signals.

The supplementary control abilities of HVDC and FACTS devices are developed for the establishment of a wide-area damping control network. The research results indicate that if the suitable input signals are selected for the WACN including multiple WADCs, it can achieve effective performance of damping multiple inter-area oscillations. The results also indicate that although there is no additional robust design for the multiple WADCs, the closed-loop system with WADCs still maintains robustness at different operating conditions, which also verify the effectiveness of the proposed signal selection method. The proposed method is useful for the selection of feedback control-inputs of multiple WADCs, which is practical for wide-area stability analysis and control of large interconnected systems.

References

1. Kundur P (1994) Power system stability and control. McGraw-Hill, New York
2. Kunjumuhammed LP, Singh R, Pal BC (2010) Probability based control signal selection for inter area oscillations damping using TCSC. In: Proceedings of IEEE power and energy society general meeting, 25–29 July 2010
3. Sadikovic R, Korba P, Andersson G (2005) Application of FACTS devices for damping of power system oscillations. In: Proceedings of IEEE PowerTech, 27–30 June 2005
4. Heniche A, Kamwa I (2008) Assessment of two methods to select wide-area signals for power system damping control. IEEE Trans Power Syst 23(2):572–581
5. Farsangi MM, Song YH, Lee KY (2004) Choice of FACTS device control inputs for damping inter-area oscillations. IEEE Trans Power Syst 19(2):1135–1143
6. Farsangi MM, Nezamabadi-pour H, Song YH, Lee KY (2007) Placement of SVCs and selection of stabilizing signals in power systems. IEEE Trans Power Syst 22(3):1061–1071
7. Skogestad S, Postethwaite I (1996) Multivariable feedback control, analysis and design. Wiley, New York
8. Milanovic JV, Duque ACS (2004) Identification of electromechanical modes and placement of PSSs using relative gain array. IEEE Trans Power Syst 19(1):410–417
9. Li Y, Rehtanz C, Yang DC, Ruberg S, Hager U (2011) Robust high-voltage direct current stabilising control using wide-area measurement and taking transmission time delay into consideration. IET Gener Transm Distrib 5(3):289–297
10. Nguyen-Duc H, Dessaint L, Okou AF, Kamwa I (2010) A power oscillation damping control scheme based on Bang-Bang modulation of FACTS signals. IEEE Trans Power Syst 25 (4):1918–1927
11. Ray S, Venayagamoorthy GK (2008) Wide-area signal-based optimal neurocontroller for a UPFC. IEEE Trans Power Deliv 23(3):1597–1605

Chapter 10
Free-Weighting Matrix Method for Delay Compensation of Wide-Area Signals

In this chapter, in order to improve the power system damping and robustness for the FACTS device, a free-weighting matrices (FWMs) approach-based Lyapunov functional stability theory is proposed to design the FACTS-WADC, which can consider efficiently the effect of signal delay on the control performance. The FWMs approach will be described and the detailed nonlinear simulations on two typical test systems will be performed to evaluate the performance of the proposed SVC-type FACTS-WADC.

10.1 Time-Delay Power System

Delay affects wide-area signals on the control performance of the wide-area controller. For wide-area control, input signals need to be received from remote regions and, furthermore, wide-area control output signals have to be sent to other remote regions where the controllable devices are located. During signal processing, a time delay is inevitably caused. Although the advanced synchronized phasor technology can control the time delay in around 300 ms [1, 2], many research results have shown that even a small time delay can still lead to control failure of the wide-area controller. More seriously, under the effect of time-delay of wide-area signals, the wide-area control may even destroy the stability of power systems. Therefore, an advanced control theory or time-delay compensation method should be considered to reduce or suppress the delay effect. At present, many controller design methods [3–7] have been proposed to handle with the time delay, but most of them only concern a fixed time delay. Under the assumption of specified time delays, these methods can reach good control performance. But in practice, the time delay of wide-area signals varies with time, and in such a case, these methods are not good to handle with time-varying delay, which limits the control performance of the wide-area controller.

For WAMS, the use of wide-area signals implies a time delay caused by the communication network [8]. The delay effect not only limits the real-time measurement performance of WAMS, but also reduces the control performance of wide-area control strategies. Many design methods have been proposed considering

© Springer-Verlag Berlin Heidelberg 2016
Y. Li et al., *Interconnected Power Systems*, Power Systems,
DOI 10.1007/978-3-662-48627-6_10

the delay of wide-area control signal, such as the decentralized/hierarchical control [9], the two-level control [3], and the H_∞ control design method [6]. For these methods, the delay of wide-area signal is simplified to a constant delay, and the controllers are designed only considering the compensation for such constant delay. However, in a real WAMS, the delay is not constant but varying with time, thus, in essence, these methods cannot reduce the varying-delay effect completely.

In fact, power systems with time-varying delay can be modeled as a standard time-delay system. The advanced control theory and method can be used to deal with the varying-delay effect of wide-area signals.

10.1.1 Description of Delay Power System with Wide-Area Signals' Delay

Wide-area closed-loop control systems use remote signals as feedback input signals. The dependence of the control system on wide-area signals makes the time delay involved in their transmission of significant concern. The time delay of the data transmission can vary from tens to hundreds of milliseconds. It depends on the communication distance, protocols, and time consumed by numerical calculations [9]. In this section, the effect of time delay in the wide-area damping control system is tested.

The delays involved in the transmission of a wide-area signal can be defined as follows, with reference to Fig. 10.1. A packet of measurements (real-time oscillatory signal) produced by a PMU is tagged with a timestamp defined as 'T1', and then the measurements are transmitted to the WAMCS center. With this real-time oscillatory signal, the WAMCS center calculates the corresponding commands for the HVDC converter control and sends these commands to the remote HVDC rectifier station. The time when the execution units in the HVDC converter stations

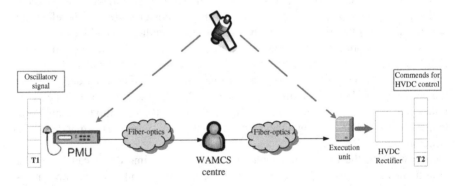

Fig. 10.1 An illustration of the time delay involved in the data transmission in GB WAMCS

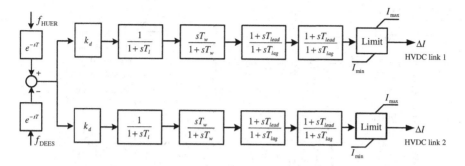

Fig. 10.2 A block diagram of HVDC damping controllers (with time delay)

receives these commands is defined as 'T2'. The difference between 'T2' and 'T1' is defined as the time delay of the data transmission in the wide-area control system.

Figure 10.2 shows a block diagram of HVDC damping controllers, in which the time delay of the transmission of the input signals is taken into account. The element e^{-sT} represents the effect of time delay; T represents the time delay in seconds [10]. In these tests, the time delays in different signal transmission channels are assumed to be constant.

Figure 10.3 presents the effects of different time delays in the wide-area control system, ranging from 50 to 200 ms. The effect of wide-area HVDC control was gradually reduced when the delay time increased. In addition, when the time delay increased to 150 ms, a high frequency oscillatory component appeared [10].

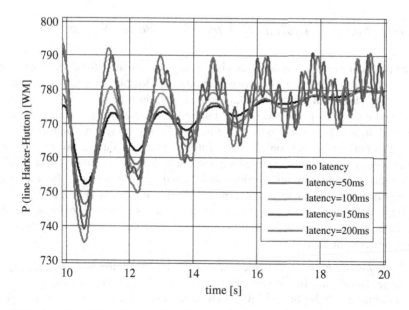

Fig. 10.3 The effect of time delay in the wide area damping controllers (50–200 ms)

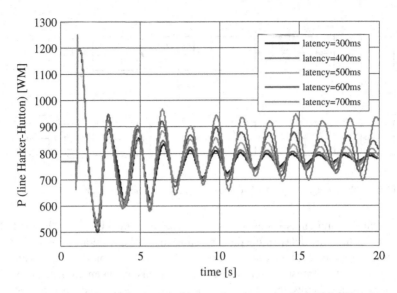

Fig. 10.4 The effect of time delay in the wide area damping controllers (300–700 ms)

Figure 10.4 presents the effect of the time delay in wide-area control system, from 300 to 700 ms. As seen from the simulations, when the time delay increased to 300 ms, the high frequency oscillatory component disappeared [10] and the inter-area oscillation became unstable as the time delay approached 700 ms.

10.1.2 Stability Analysis of Time-Delay Power System

In the physical world, the system trend is not only dependent on the current operation point but is also subject to its previous conditions. This phenomenon is denoted as time delay [11]. Although time delay exists widely in the power system measurement and control loops, it was usually ignored in the past stability study and controller design. The reason was that traditional power system controllers were generally designed based on the local information; and time delay of the local measurement was small enough to be ignored with little error. However, time delay is significant (from tens milliseconds to hundreds milliseconds) in the wide-area environment and cannot be simply ignored. So it is important to properly evaluate the impact of time delay on power system stability assessment and controller design.

There have been some stability analysis methods for time-delay system, including but not limited to:

- *Time domain method.* This method directly integrates the system trajectory to judge its stability. In the calculation, some transformations are usually adopted to approximate the time delays, such as rational approximation method, Pade approximation method, etc. [11–13].

- *Eigenvalue-based method.* In this type of method, total or partial eigenvalues of the time-delay system are calculated to evaluate its stability. Smith predictor method [14], Lambert-W function method [15], root clustering paradigm method [16] can be regarded as this type.
- *Lyapunov method* [17–21]. Such method evaluates system stability by constructing various Lyapunov function/ functional. The stability criterion can be classified into two types: delay-dependent and delay independent. The former is usually considered to have less conservativeness, especially when the delay is small.

Time domain method and eigenvalue-based method can only be suitable for the analysis of time invariant delay system, while Lyapunov method has much more applicability and flexibility. It can be applied to the stability analysis of time variant, nonlinear, switched, and uncertain time-delay systems. The key of this method is to find more suitable Lyapunov function and reduce its conservativeness.

10.2   Description of Free-Weighting Matrices (FWMs) Method

Research of the stability of time-delay systems began in the 1950s, including frequency-domain approaches and time domain methods. Figure 10.5 gives an overall picture of the research on the stability of time-delay systems, such as

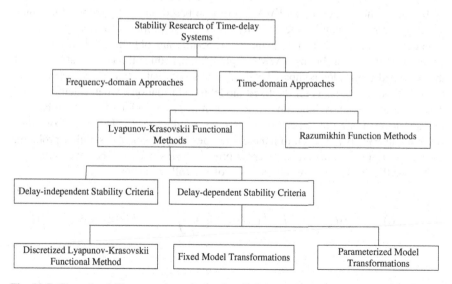

Fig. 10.5 General stability research methods of typical time-delay system

$$\begin{cases} \dot{x}(t) = Ax(t) + Bx(t-h) + Cu(t) \\ x(t) = \phi(t) \end{cases} \tag{10.1}$$

in which A, B, and C are the constant real matrices with appropriate dimensions. h is the time delay. $u(t)$ is the input single. $\phi(t)$ is the initial condition.

As for frequency-domain approaches, it determines the stability of a system from the distribution of the roots of its characteristic equation or from the solutions of a complex Lyapunov matrix function equation. If and only if $f(\lambda) = \det(\lambda I - A - Be^{-h\lambda}) = 0$ have negative real part, the time-delay system (10.1) is stable. Since this equation is transcendental, it is difficult to solve. So the time domain methods are very important in the stability analysis of linear systems. Among them, two classes of sufficient conditions have received a lot of attention: delay-dependent criteria and delay-independent criteria. Since the delay-dependent stability criteria make use of information on the size of delay, while the delay-independent stability criteria do not include such information, the delay-dependent conditions are generally less conservative than the delay-independent ones, especially when the time delays are small.

So far, three methods of studying delay-dependent problems have been devised: the discretized Lyapunov–Krasovskii functional method, fixed model transformations, and parameterized model transformations. The discretized Lyapunov functional method [16–18] is one of the most efficient among them, but it is difficult to extend to the synthesis of a control system. Another method involves a fixed model transformation, which expresses the delay term in terms of an integral. Four basic model transformations have been proposed [19]. Combined with Park's or Moon et al.'s inequalities [20], the descriptor model transformation method is the most efficient [19, 21, 22]. However, there is room for further investigation.

Recently, they proposed an FWMs approach based on the Lyapunov functional stability theory to handle the stability/stabilization problem for many kinds of time-delay system and many advantageous results are obtained [23–25]. For the controllers' design problems, FWMs approach can reduce the conservativeness of the designed controller compared with the conventional fixed model transformations methods. That is, as for the fixed model transformations, when calculating the derivative of Lyapunov function, some inequalities such as Park and Moon have to be used to estimate the upper bound of cross-product terms. Different from this, when using FWMs approach to handle with the delay-dependent stability problem, the bounding techniques on some cross product terms need not be involved.

The well-used Lyapunov functional is of the following form:

$$V(t, x_t) = x^T(t)Px(t) + \int_{t-h}^{t} x^T(s)Qx(s)\mathrm{d}s + \int_{-h}^{0} \int_{t+\theta}^{t} \dot{x}^T(s)Z\dot{x}(s)\mathrm{d}s\mathrm{d}\theta \tag{10.2}$$

where the first and second terms of Lyapunov functional can be viewed as things like the kinetic and potential energy of a mechanical system, and the third term is introduced to make the derivation simple. $P = P^T, Q = Q^T, Z = Z^T$ are to be determined. Then calculate the derivative of Lyapunov functional in (10.2):

$$\dot{V}(x_t) = \dot{x}^T(t)Px(t) + x^T(t)P\dot{x}(t) + x^T(t)Qx(t)$$

$$- x^T(t-h)Qx(t-h) + h\dot{x}^T(t)Z\dot{x}(t) - \int_{t-h}^{t} \dot{x}^T(s)Z\dot{x}(s)ds \qquad (10.3)$$

In order to obtain the LMIs-based stability criteria, by employing the conventional fixed model transformations and Lyapunov stability theorem, it is usually added to the right side of the following terms into the derivative of $V(t, x_t)$ in (10.3):

$$0 = 2x^T(t)PB\left[x(t) - x(t-\tau) - \int_{t-\tau}^{t} \dot{x}(s)ds\right] \qquad (10.4)$$

where B is coefficient matrix and P is Lyapunov matrix. From this, it is obvious to see that as for the conventional solutions such as fixed model transformations, when we solve the inequalities that ensure $\dot{V}(x_t) < 0$, B and P have to be adjusted and they cannot be selected freely, which is a serious limitation for conventional solutions.

However, in order to reduce the conservativeness, when we dealt with the derivative of Lyapunov functional in (10.2), we employed two new free-weighting matrices M and N, to explain the relationship between the New-Leibniz formula instead of P and B, which can be called the free-weighting matrices (FWMs) approach.

So in the process of derivative calculation of (10.3), it equals to adding the following right sides of (10.5) into $\dot{V}(x_t)$, then we can also obtain some LMIs inequalities to ensure $\dot{V}(x_t) < 0$. From this, it can be seen that for FWMs approach, it need not select B and P, but only optimize M and N by solving LMIs. Hence, the FWMs approach can reduce the conservativeness of the controller to the time-delay system. In Sect. 10.4, we describe the solving process by FWMs approach in detail.

$$0 = \left[x^T(t)M + \dot{x}^T(t)N\right]\left[x(t) - x(t-\tau) - \int_{t-\tau}^{t} \dot{x}(s)ds\right] \qquad (10.5)$$

10.3   General Configuration of FACTS-WADC Based on FWMs Approach

The concept on FACTS-WADC can be described through Fig. 10.6, which shows one power systems installed with SVC-type FACTS device and its supplementary wide-area robust damping controller. As Fig. 10.2 shows, the closed-loop feedback control formed by SVC-type FACTS-WADC can be used to damp the potential power oscillations. For the wide-area robust damping controller, its control-input signals select the operating variables from global range of power systems, and such selection can be performed based on small signal analysis (SSA) or other methods [26]. In addition, the application of WAMS for the transmission of the wide-area control signals inevitably causes time delay, thus, a power system with WAMS application is in essence a typical time-delay system. In this chapter, the first-order Pade approximation is used to model the time delay characteristic of the wide-area control signals. Meanwhile, the high-pass and the low-pass filters (HPF and LPF) are necessary to process the control-input signals with the purpose of retaining the concerned oscillation frequency information.

Furthermore, for the robust wide-area FACTS damping controller, the extremely important part is the state-feedback control gain matrix K, which can be optimally designed by employing the proposed FWMs approach. In addition, the designed K is the gain matrix for the state variables. However, the operating state variables cannot be completely observed very well for the practical system, thus, the state observer $O(s)$ is introduced to implement the observation on the state variable. In this chapter, the practical pole-placement method is used to design the $O(s)$, and the structure of $O(s)$ will be also given in the following section.

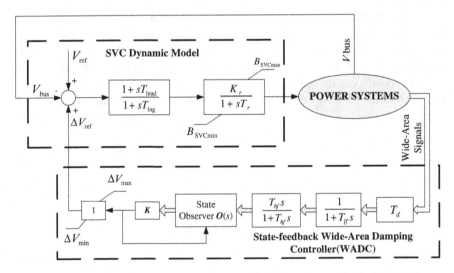

Fig. 10.6 Control block diagram of SVC-type FACTS device with state-feedback WADC

10.4 FWMs Approach-Based FACTS-WADC Design

In Chap. 6, the power system model, which includes generators, loads, various controllers, FACTS devices, etc., is described using a set of differential-algebraic equations [27, 28]. In this way, the open-loop system model can be modeled, which uses the output of the FACTS supplementary control as the input and while the wide-area control signals as the output of the open-loop power system. Furthermore, when the time-delay τ is considered as the transmission delay of the control input of the FACTS wide-area robust damping controller, the general linearized power system model can be modified as follows:

$$\begin{cases} \dot{x}(t) = Ax(t) + Bu(t) \\ y(t) = Cx(t) \end{cases} \tag{10.6}$$

where A is the state matrix, B is the input matrix, and C is the output matrix.

The state-feedback controller that will be designed can be simplified as follows:

$$u(t) = Kx(t - \tau) \tag{10.7}$$

Then the closed-loop model constructed by open-loop model (10.6) and FACTS-WADC (10.7) can be written as

$$\begin{cases} \dot{x}(t) = Ax(t) + BKx(t - \tau) \\ y(t) = Cx(t) \end{cases} \tag{10.8}$$

The objective of this section is to develop a new delay-dependent stabilization criterion that provides the optimal controller gain K and time-delay h ($h = \max(\tau)$), such that the resulting closed-loop system (10.8) is asymptotically stable. For this purpose, the following lemmas will be employed in the proofs of our results. The notion * stands for the symmetric matrix.

Lemma (Schur complement) [29]: *Given a symmetric matrix* $S = S^T =$ $\begin{bmatrix} S_{11} & S_{12} \\ S_{12}^T & S_{22} \end{bmatrix}$ ($S_{11} \in \Re^{r \times r}$), *the following three conditions are equal:*

(1) $S < 0$
(2) $S_{11} < 0$, $S_{22} - S_{12}^T S_{11}^{-1} S_{12} < 0$
(3) $S_{22} < 0$, $S_{11} - S_{12} S_{22}^{-1} S_{12}^T < 0$

Theorem *For given scalar h, there exists a state-feedback controller of (10.7) such that the closed-loop system (10.8) is asymptotically stable if there exist* $L = L^T > 0$,

$Q_1 = Q_1^T > 0$, $R = R^T > 0$, $Y = \begin{bmatrix} Y_{11} & Y_{12} \\ Y_{12}^T & Y_{22} \end{bmatrix} \geq 0$ *and any appropriately* *dimensioned matrices M_1, M_2 and V, such that the following matrix inequalities are feasible*:

$$\bar{\Phi} = \begin{bmatrix} AL + LA^T + M_1 + M_1^T + Q_1 + hY_{11} & BV - M_1 + M_2^T + hY_{12} & hLA^T \\ * & -M_2 - M_2^T - Q_1 + hY_{22} & hV^T B^T \\ * & * & -hR \end{bmatrix} < 0 \tag{10.9}$$

$$\bar{\Psi} = \begin{bmatrix} Y_{11} & Y_{12} & M_1 \\ * & * & M_2 \\ * & * & LR^{-1}L \end{bmatrix} > 0 \tag{10.10}$$

Moreover, the feedback controller gain is $K = VL^{-1}$.

Proof It is clear from Newton–Leibniz formula that

$$x(t) - x(t - \tau) - \int_{t-\tau}^{t} \dot{x}(s)\mathrm{d}s = 0 \tag{10.11}$$

Hence from (10.11), for any appropriately dimensioned matrices N_1 and N_2, the following equation is true:

$$0 = 2\left(x^T(t)N_1 + x^T(t - \tau)N_2\right)\left[x(t) - x(t - \tau) - \int_{t-\tau}^{t} \dot{x}(s)\mathrm{d}s\right] \tag{10.12}$$

On the other hand, for any semi-positive definite matrix $X = \begin{bmatrix} X_{11} & X_{12} \\ X_{12}^T & X_{22} \end{bmatrix} \geq 0$, the following equation holds:

$$h\xi^T(t)X\xi(t) - \int_{t-\tau}^{t} \xi^T(t)X\xi(t)\mathrm{d}s > 0 \tag{10.13}$$

where $\xi(t) = [x^T(t),\, x^T(t - \tau)]^T$.

Construct the following Lyapunov candidate function:

$$V(x_t) = x^T(t)Px(t) + \int_{t-\tau}^{t} x^T(s)Qx(s)\mathrm{d}s + \int_{-h}^{0} \int_{t+\theta}^{t} \dot{x}^T(s)Z\dot{x}(s)\mathrm{d}s\mathrm{d}\theta \tag{10.14}$$

where $P = P^T > 0$, $Q = Q^T > 0$, $Z = Z^T > 0$ are to be determined.

Calculating the derivatives of $V(x_t)$ defined in (10.14) along the trajectories of system (10.8) yields

$$\dot{V}(x_t) = \dot{x}^T(t)Px(t) + x^T(t)P\dot{x}(t) + x^T(t)Qx(t)$$

$$- x^T(t-\tau)Qx(t-\tau) + h\dot{x}^T(t)Z\dot{x}(t) - \int_{t-h}^{t} \dot{x}^T(s)Z\dot{x}(s)ds$$

$$= x^T(t)[PA + A^TP]x(t) + 2x^T(t)PBKx(t-\tau) + x^T(t)Qx(t) - x^T(t-\tau)Qx(t-\tau)$$

$$+ h[Ax(t) + BKx(t-\tau)]^T Z[Ax(t) + BKx(t-\tau)]^T - \int_{t-h}^{t} \dot{x}^T(s)Z\dot{x}(s)ds$$

$$= x^T(t)[PA + A^TP + hA^TZA + Q + hX_{11} + N_1 + N_1^T]x(t)$$
$$+ x^T(t)[PBK + hA^TZBK - N_1 + N_2^T + hX_{12}]x(t-\tau)$$
$$+ x^T(t-\tau)[PBK + hA^TZBK - N_1 + N_2^T + hX_{12}]^T x(t)$$
$$+ x^T(t-\tau)[-N_2 - N_2^T - Q + hX_{22} + hK^TB^TZBK]^T x(t-\tau)$$

$$- \int_{t-\tau}^{t} [\dot{x}^T(s)Z\dot{x}(s) + \xi^T(t)X\xi(t) + 2(x^T(t)N_1 + x^T(t-\tau)N_2)\dot{x}(s)]ds$$

$$(10.15)$$

Then adding the terms on the right of (10.12) and (10.13) to $\dot{V}(x_t)$, it becomes

$$\dot{V}(x_t) \leq x^T(t)[PA + A^TP]x(t) + 2x^T(t)PBKx(t-\tau) + x^T(t)Qx(t) - x^T(t-\tau)Qx(t-\tau)$$

$$+ h[Ax(t) + BKx(t-\tau)]^T Z[Ax(t) + BKx(t-\tau)] - \int_{t-\tau}^{t} \dot{x}^T(s)Z\dot{x}(s)ds + h\xi^T(t)X\xi(t)$$

$$- \int_{t-\tau}^{t} \xi^T(t)X\xi(t)ds + 2(x^T(t)N_1 + x^T(t-\tau)N_2)\left[x(t) - x(t-\tau) - \int_{t-\tau}^{t} \dot{x}(s)ds\right]$$

$$= x^T(t)[PA + A^TP]x(t) + 2x^T(t)PBKx(t-\tau) + x^T(t)Qx(t) - x^T(t-\tau)Qx(t-\tau)$$
$$+ h\xi^T(t)X\xi(t) + h[Ax(t) + BKx(t-\tau)]^T Z[Ax(t) + BKx(t-\tau)]$$

$$+ 2(x^T(t)N_1 + x^T(t-\tau)N_2)[x(t) - x(t-\tau)] - \int_{t-\tau}^{t} \dot{x}^T(s)Z\dot{x}(s)ds$$

$$- \int_{t-\tau}^{t} \xi^T(t)X\xi(t)ds - 2(x^T(t)N_1 + x^T(t-\tau)N_2)\int_{t-\tau}^{t} \dot{x}(s)ds$$

$$(10.16a)$$

$$
\begin{aligned}
\dot{V}(x_t) \leq\ & x^T(t)\left[PA + A^TP + hA^TZ\Lambda + Q + hX_{11} + N_1 + N_1^T\right]x(t) \\
& + x^T(t)\left[PBK + hA^TZBK - N_1 + N_2^T + hX_{12}\right]x(t-\tau) \\
& + x^T(t-\tau)\left[PBK + hA^TZBK - N_1 + N_2^T + hX_{12}\right]^T x(t) \\
& + x^T(t-\tau)\left[-N_2 + N_2^T - Q + hX_{22} + hK^TB^TZBK\right]^T x(t-\tau) \\
& - \int_{t-\tau}^{t}\left[\dot{x}^T(s)Z\dot{x}(s) + \xi^T(t)X\xi(t) + 2\left(x^T(t)N_1 + x^T(t-\tau)N_2\right)\dot{x}(s)\right]ds
\end{aligned}
$$

$$
= \begin{bmatrix} x(t) \\ x(t-\tau) \end{bmatrix}^T
\begin{bmatrix}
PA + A^TP + Q + hX_{11} + & PBK + hA^TZBK - \\
N_1 + N_1^T + hA^TZA & N_1 + N_2^T + hX_{12} \\
& -N_2 - N_2^T - Q + \\
* & hX_{22} + hK^TB^TZBK
\end{bmatrix}
\begin{bmatrix} x(t) \\ x(t-\tau) \end{bmatrix}
$$

$$
- \int_{t-\tau}^{t}
\begin{bmatrix} x(t) \\ x(t-\tau) \\ \dot{x}(s) \end{bmatrix}^T
\begin{bmatrix}
X_{11} & X_{12} & N_1 \\
* & X_{22} & N_2 \\
* & * & Z
\end{bmatrix}
\begin{bmatrix} x(t) \\ x(t-\tau) \\ \dot{x}(s) \end{bmatrix} ds
\tag{10.16b}
$$

Define

$$
\Xi =
\begin{bmatrix}
PA + A^TP + Q + hX_{11} + & PBK + hA^TZBK - \\
N_1 + N_1^T + hA^TZA & N_1 + N_2^T + hX_{12} \\
& -N_2 - N_2^T - Q + \\
* & hX_{22} + hK^TB^TZBK
\end{bmatrix}
\tag{10.17}
$$

$$
\Psi =
\begin{bmatrix}
X_{11} & X_{12} & N_1 \\
* & X_{22} & N_2 \\
* & * & Z
\end{bmatrix}
\tag{10.18}
$$

From (10.16a, 10.16b), we can see if $\Xi < 0$ and $\Psi > 0$, then $\dot{V}(x_t) < 0$. This means that the closed-loop system (10.8) is asymptotically stable. Using **Lemma**, it is easy to obtain $\Xi < 0$ which is equal to

$$
\Phi =
\begin{bmatrix}
PA + A^TP + Q + hX_{11} + N_1 + N_1^T & PBK - N_1 + N_2^T + hX_{12} & hA^TZ \\
* & -N_2 - N_2^T - Q + hX_{22} & hK^TB^TZ \\
* & * & -hZ
\end{bmatrix} < 0
\tag{10.19}
$$

In order to solve the controller gain K, define

$$L = P^{-1}, \quad M_1 = P^{-1}N_1P^{-1}, \quad M_2 = P^{-1}N_2P^{-1}$$
$$R = Z^{-1}, \quad V = KP^{-1}, \quad Q_1 = P^{-1}QP^{-1}$$
$$Y = \text{diag}\{P^{-1}, P^{-1}\}X\text{diag}\{P^{-1}, P^{-1}\}$$

Pre- and post-multiply the left and right sides of (10.19) by $\text{diag}\{P^{-1}, P^{-1}, Z^{-1}\}$, respectively, that is,

$$\begin{bmatrix} P^{-1} & 0 & 0 \\ 0 & P^{-1} & 0 \\ 0 & 0 & Z^{-1} \end{bmatrix} \begin{bmatrix} PA + A^TP + Q + & PBK - N_1 + & \\ hX_{11} + N_1 + N_1^T & N_2^T + hX_{12} & hA^TZ \\ & -N_2 - N_2^T - & \\ * & Q + hX_{22} & hK^TB^TZ \\ * & * & -hZ \end{bmatrix} \begin{bmatrix} P^{-1} & 0 & 0 \\ 0 & P^{-1} & 0 \\ 0 & 0 & Z^{-1} \end{bmatrix}$$

$$= \begin{bmatrix} AL + LA^T + M_1 + M_1^T + Q_1 + hY_{11} & BV - M_1 + M_2^T + hY_{12} & hLA^T \\ * & -M_2 - M_2^T - Q_1 + hY_{22} & hV^TB^T \\ * & * & -hR \end{bmatrix} < 0$$

(10.20)

It is clear that (10.19) becomes (10.9).

Similarly, pre- and post-multiply the left and right sides of (10.18) by $\text{diag}\{P^{-1}, P^{-1}, P^{-1}\}$, respectively, (10.18) becomes (10.10). This completes the proof of theorem.

Since the conditions in theorem are no longer LMIs owing to the nonlinear terms $LR^{-1}L$ in (10.10), we cannot use a convex optimization algorithm to find a minimum value. However, we can use the cone complementarity linearization algorithm [30], which is based on LMIs, to solve this problem. The nonlinear optimization object and constraints can be constructed as follows.

$$\text{Minimize } \text{tr}\{FF_1 + LL_1 + RR_1\} \tag{10.21}$$

Subject to:

$$\begin{bmatrix} AL + LA^T + M_1 + M_1^T + Q_1 + hY_{11} & BV - M_1 + M_2^T + hY_{12} & hLA^T \\ * & -M_2 - M_2^T - Q_1 + hY_{22} & hV^TB^T \\ * & * & -hR \end{bmatrix} < 0$$

$$\begin{bmatrix} Y_{11} & Y_{12} & M_1 \\ * & Y_{22} & M_2 \\ * & * & F \end{bmatrix} > 0 \tag{10.22}$$

$$\begin{cases} \begin{bmatrix} F & I \\ I & F_1 \end{bmatrix} > 0, \begin{bmatrix} F_1 & L_1 \\ L_1 & R_1 \end{bmatrix} > 0 \\ L > 0, \ F > 0, \ R > 0 \end{cases}$$

Based on the solution of the above nonlinear optimization object, the optimal controller gain matrix K and the maximum time delay $h = \max(\tau)$ can be searched out by the following proposed iteration algorithm.

Step 1: Choose a small initial time-delay h to ensure that there exists a feasible solution for (10.9), (10.21), and (10.22).

Step 2: Initially set a set of feasible matrix variable values for (L', L_1', V', M_1', M_2', F', F_1', Q_1', R', R_1', Y'), which should satisfy (10.9), (10.21), and (10.22). Then, set $k = 0$.

Step 3: Solve the above nonlinear optimization problem expressed by the LMI constraints (10.9), (10.21), and (10.22). Then, set $F_{k+1} = F$, $F_{1,k+1} = F_1$, $L_{k+1} = L, L_{1,k+1} = L_1, R_{k+1} = R, R_{1,k+1} = R_1$.

Step 4: If inequality (10.10) is feasible, then increase h by a small amount and return to *Step 2*. If inequality (10.10) is unfeasible within a specified number of iterations, then stop. Otherwise, set $k = k+1$ and go to *Step 3*.

According to the theorem and algorithm, we can obtain the optimized feedback controller gain K and the maximum time delay h. In addition, it is necessary to remark that in the practical power system, since the operating state variables cannot be completely observed, it is preferable to use feedback control with measurable states for practical applications. For this purpose, the state observer $O(s)$ should be introduced to implement the observation on the state variable transmitted by WAMS. In this chapter, the state observer is designed with the conventional pole-placement method [31]. Finally, the designed SVC-type FACTS wide-area robust damping control, which considers the time delay of wide-area measurement signals, can be expressed as Fig. 10.7 shows.

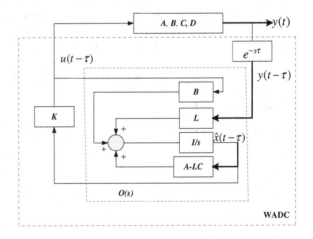

Fig. 10.7 Configuration of SVC-type FACTS-WADC including state observer

10.5   Cases Study

10.5.1   4-Machine 2-Area System

In this case, the 4-machine 2-area benchmark system [27] for inter-area oscillation study is used to validate the proposed robust FACTS-WADC. Figure 10.8 shows the system modified with a shunt FACTS device on bus-8. Such FACTS device is mainly used to improve the voltage profile and interconnected ability of the areas. In this system, G1 and G2 are located in area-1, and G3 and G4 are in area-2.

Many research results have demonstrated that there is a typical inter-area oscillation mode between area-1 and area-2, and such oscillation further represents the power oscillation between G1 and G3. Thus, in this case, the rotor angle and speed deviation of the both machines are selected as the wide-area feedback input for the FACTS device. In addition, the classic shunt FACTS device, that is SVC, is used to implement the supplementary wide-area control.

The nonlinear dynamical simulation is performed by set three-phase ground-fault on bus-8 for the duration of 190 ms. Figure 10.9 shows the response of line power flow from area-1 to area-2. From this, it can be seen that as for the system without the SVC wide-area damping control, such large disturbance leads to serious power oscillation occurrence in the interconnected system. However, when the presented control is applied, such oscillation is damped effectively, there is 180 ms delay time for the wide-area control signals, and the designed controller still maintains good damping effects. Figure 10.10 shows the relative angular between G1 and G3. From this, it can be seen that there is oscillation on one generator in one area against another generator in another area, and such oscillation is the direct reason for power oscillation of the interconnected system. However, with implementation of the SVC wide-area control, such oscillation is damped very well, which further verifies the correctness of the designed controller.

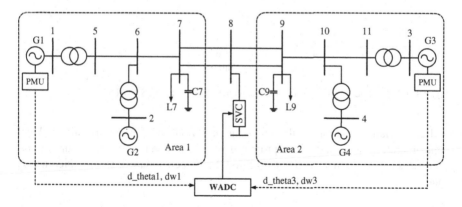

Fig. 10.8 The 4-machine 2-area system

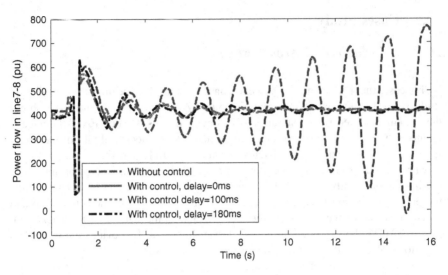

Fig. 10.9 Power flow from area-1 to area-2

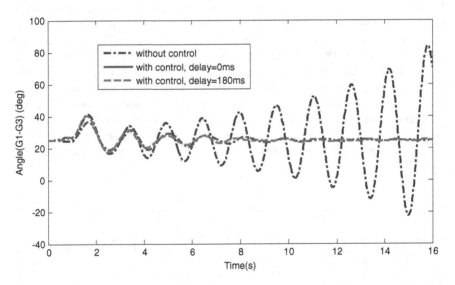

Fig. 10.10 Dynamic response of the system with SVC-type FACTS-WADC

Furthermore, to compare the damping performance of the conventional local control strategy with that of the proposed wide-area control strategy, four simulation cases are performed on the 4-machine 2-area benchmark system, which are as follows:

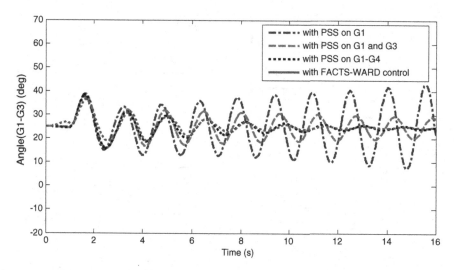

Fig. 10.11 Dynamic response of the system with local PSS or with SVC-type FACTS-WARD controll

- Case-1: One general local power system stabilizer (PSS) is installed on G1 located in Area-1;
- Case-2: Two local PSSs are installed on G1 (in Area-1) and G3 (in Area-2);
- Case-3: All the generators in the two areas are installed with the local PSSs;
- Case-4: Do not install any local PSS, and only perform the FACTS wide-area robust damping (WARD) control.

Figure 10.11 shows the results of the above four simulation cases. From this, it can be clearly seen that when it just implements the local control strategy, the single or part local PSSs cannot provide sufficient damping on the inter-area oscillation occurrence between different areas. Although when all the generators install the local PSSs, it can achieve the accepted damping performance like the implementation of the FACTS-WARD control; it should be noted that the coordination among plenty of PSSs in different areas could be difficult in practical large-scale power systems. However, as for the FACTS-WARD, there is no related problem and it just introduces the supplementary control of FACTS device to construct the wide-area damping control strategy to prevent inter-area oscillations.

10.5.2   16-Machine 5-Area Test System

As Fig. 10.12 shows, the 16-machine 5-area test system [17] modified with a shunt FACTS device on bus-51 is simulated in this chapter to validate the proposed FWMs-based robust controller design approach for FACTS wide-area damping

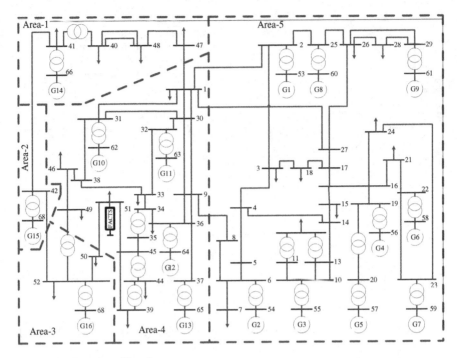

Fig. 10.12 The 16-machine 5-area test system

control. This is in fact the simplified New England and New York interconnected system. So, the first nine machines (G1–G9) and the second four machines (G10–G13) belong to the New England Test System (NETS) and the New York Power System (NYPS), respectively. In addition, there are three more machines (G14–G16) used as the dynamical equivalent of the three neighbor areas connected with NYPS area. It should be remarked that all the machines are described by the sixth-order dynamical model.

As the classic shunt-type FACTS device, SVC is installed on the interconnected bus to improve the voltage profile of the system. The mode analysis on the test system shows that there is an inter-area oscillation mode (oscillation frequency is 0.57 Hz and damping ratio is 0.017) between G14 and G16 under the specific operating condition. As for such low-frequency oscillator (LFO) mode, the SVC wide-area supplementary control strategy could be a better alternative to the conventional local PSS control strategy. However, to implement the effective wide-area damping control, it is important to select the suitable wide-area feedback signals. The residue analysis results show that the inter-area mode represents high

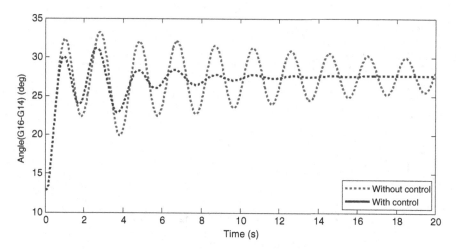

Fig. 10.13 Dynamic response of the system with SVC-type FACTS-WADC

observability on the real line current between bus-68 and bus-52. Thus, such line current is selected as the feedback stabilizing signal for SVC wide-area damping control. Furthermore, based on the linearized model of the test system, robust control design can be performed step-by-step as the above proposed method.

A three-phase short-circuit fault with 100 ms duration is applied nearby the bus-51 to assess the control performance under large disturbance. Figures 10.13, 10.14, 10.15, 10.16, 10.17, 10.18 show the nonlinear simulation results on the system dynamic response. Figure 10.13 shows the relative angular between G14 and G16. From this, it can be seen that low-frequency oscillation between G14 and G16 is damped very well by the presented SVC-type FACTS-WADC. Meanwhile, the power oscillations about the power flow nearby the machine and at the interconnected backbone line are also damped significantly as Fig. 10.14 shows. The output variation of the SVC shown in Fig. 10.15 indicates that the SVC with wide-area supplementary control takes quick action against oscillation damping.

To further reveal the control performance of the presented robust controller under different delays of wide-area control signal, the 100 ms and the 200 ms delay caused by signal transmission is simulated respectively. Figures 10.16, 10.17, 10.18 show comparison of the dynamic response under such different delays. From these, it can be seen that with the increase in the time delay from 0 to 200 ms, the designed controller always maintains good oscillation damping effect, which indicates the robustness of the presented controller against the transmission delays of the wide-area feedback signal.

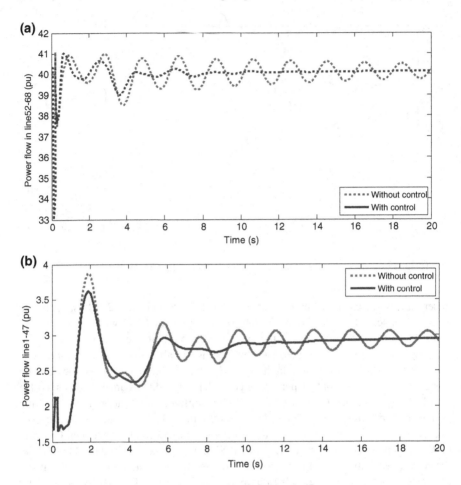

Fig. 10.14 Power flow in the tie-line. **a** Line 52–68. **b** Line 1–47

10.6 Summary

This chapter presents a new linear design approach on the robust FACTS wide-area damping control to enhance the power stability of large power systems. FWMs are introduced to convert the optimization object with nonlinear matrix inequality constraints into a set of LMI constraints, which is convenient to the linear design for the FACTS wide-area robust damping controller. Furthermore, a nonlinear optimization algorithm is presented to search out the optimal control gain matrix with the maximum delay independent of the wide-area feedback control signals. Thus, it

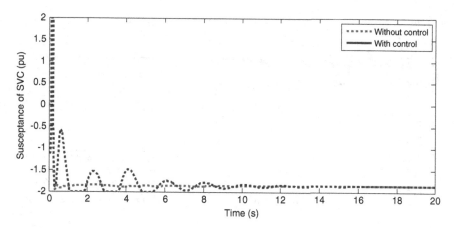

Fig. 10.15 Output of the SVC

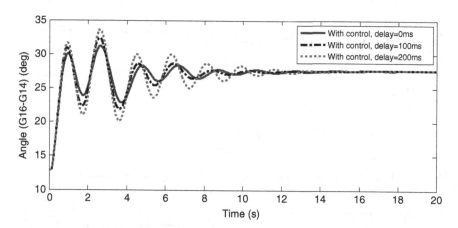

Fig. 10.16 Dynamic response of the system with SVC-type FACTS-WADC under different time delays

can effectively improve the negative effects for the time-varying delays on the control performance, which is also in practice in large power networks with wide-area control system. In practice, after the acceptable system equivalent, the proposed method is simple and easy to be applied in large power systems. Nonlinear simulations on both the 4-machine 2-area and the 16-machine 5-area system show that the presented robust FACTS wide-area controller not only improves the system oscillation stability, but also has robustness against the variations of time delay aroused by the wide-area transmission and processing information in WAMS.

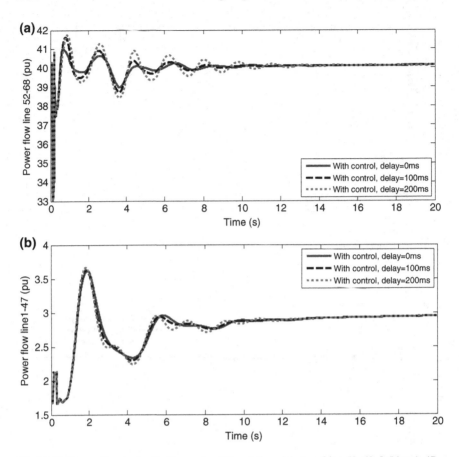

Fig. 10.17 Power flow in the tie-line under different time delays. a Line 52–68. b Line 1–47

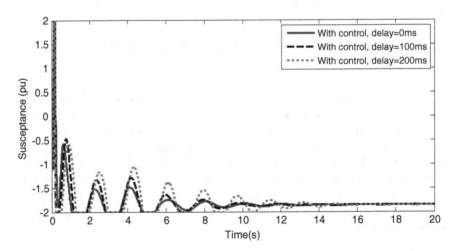

Fig. 10.18 Output of the SVC under different time delays

References

1. Wu HX, Tsakalis KS, Heydt GT (2004) Evaluation of time delay effects to wide-area power system stabilizer design. IEEE Trans Power Syst 19(4):1935–1941
2. Stahlhut JW, Browne TJ, Heydt GT, Vittal V (2008) Latency viewed as a stochastic process and its impact on wide area power system control signals. IEEE Trans Power Syst 23(1):84–91
3. Dotta D, Silva AS, Decker IC (2009) Wide-area measurement-based two-level control design considering signals transmission delay. IEEE Trans Power Syst 24(1):208–216
4. Majumder R, Pal BC, Dufour C, Korba P (2006) Design and real-time implementation of robust FACTS controller for damping inter-area oscillation. IEEE Trans Power Syst 21(2):809–816
5. Majumder R, Chaudhuri B, Pal BC (2005) A probabilistic approach to model-based adaptive control for damping of interarea oscillations. IEEE Trans Power Syst 20(1):367–374
6. Chaudhuri B, Pal BC (2004) Robust damping of multiple swing modes employing global stabilizing signals with a TCSC. IEEE Trans Power Syst 19(1):499–506
7. Chaudhuri NR, Ray S, Majumder R, Chaudhuri B (2010) A new approach to continuous latency compensation with adaptive phasor power oscillation damping controller (POD). IEEE Trans Power Syst 25(2):939–946
8. Xie XR, Xin YZ, Xiao JY, Wu JT, Han YD (2006) WAMS applications in Chinese power systems. IEEE Power Energy Mag 4(1):54–63
9. Kamwa I, Grondin R, Hebert Y (2001) Wide-area measurement based stabilizing control of large power systems-a decentralized/hierarchical approach. IEEE Trans Power Syst 16(1):136–153
10. He J, Lu C, Jin X, Li P (2008) Analysis of time delay effects on wide area damping control. In: IEEE Asia Pacific conference on circuits and systems
11. Savelli DC, Pellanda PC, Martins N, Macedo NJP, Barbosa AA, Luz GS (2007) Robust signals for the TCSC oscillation damping controllers of the Brazilian north–south interconnection considering multiple power flow scenarios and external disturbances. In: IEEE power engineering society general meeting
12. Grunbaum R, de Grijp M, Moshi V (2009) Enabling long distance AC power transmission by means of FACTS. AFRICON, Sept 2009
13. Chaudhuri NR, Ray S, Majumder R, Chaudhuri B (2009) A case study on challenges for robust wide-area phasor POD. In: IEEE power & energy society general meeting
14. Korba P, Larsson M, Chaudhuri B, Pal B, Majumder R, Sadikovic R, Andersson G (2007) Towards real-time implementation of adaptive damping controllers for FACTS devices. In: IEEE power engineering society general meeting
15. Zarghami M, Crow ML, Jagannathan S (2010) Nonlinear control of FACTS controllers for damping interarea oscillations in power systems. IEEE Trans Power Deliv 25(4):3113–3121
16. Gu K (1997) Discretized LMI set in the stability problem for linear uncertain time-delay systems. Int J Control 68(4):923–934
17. Gu K (1999) A generalized discretization scheme of Lyapunov functional in the stability problem of linear uncertain time-delay systems. Int J Robust Nonlinear Control 9(1):1–4
18. Gu K (2001) A further refinement of discretized Lyapunov functional method for the stability of time-delay systems. Int J Control 74(10):967–976
19. Fridman E, Shaked U (2003) Delay-dependent stability and H_∞ control: constant and time-varying delays. Int J Control 76(1):48–60
20. Fridman E, Shaked U (2002) On delay-dependent passivity. IEEE Trans Autom Control 47(4):664–669
21. Gao H, Wang C (2003) Comments and further results on 'a descriptor system approach to H1 control of linear time-delay systems'. IEEE Trans Autom Control 48(3):520–525
22. Han QL (2003) Stability criteria for a class of linear neutral systems with time-varying discrete and distributed delays. IMA J Math Control Inf 20(4):371–386

23. Liu F, Wu M, He Y, Zhou YC, Yokoyama R (2010) Delay-dependent robust stability analysis for interval neural networks with time-varying delay. IEEJ, accepted, 2010

24. Liu F, Wu M, He Y, Zhou YC, Yokoyama R (2008) New delay-dependent stability criteria for T-S fuzzy systems with a time-varying delay. In: Proceedings of the 17th world congress, the international federation of automatic control. Seoul, Korea

25. Liu F, Wu M, He Y, Zhou YC, Yokoyama R (2008). New delay-dependent stability analysis and stabilizing design for T-S fuzzy systems with a time-varying delay. In: 27th Chinese control conference. Kunming, China

26. Jiang QY, Zou ZY, Cao YJ (2005) Wide-area TCSC controller design in consideration of feedback signals' time delays. In: IEEE power engineering society general meeting

27. Kundur P (1994) Power stability and control. McGraw-Hill, New York

28. Rogers G (1999) Power System oscillations. Kluwer, Norwell

29. Gao H, Lam J, Wang C, Wang Y (2004) Delay-dependent output-feedback stabilization of discrete-time systems with time-varying state delay. IEEE Proc Control Theor Appl 151 (6):691–698

30. El Ghaoui L, Oustry F, AitRami M (1997) A cone complementarity linearization algorithm for static output-feedback and related problems. IEEE Trans Autom Control 42(8):1171–1176

31. Chilali M, Gahinet P, Apkarian P (1999) Robust pole placement in LMI regions. IEEE Trans Autom Control 44(12):2257–2270 (Chap. 4 references)

Chapter 11
Design and Implementation of Delay-Dependent Wide-Area Damping Control for Stability Enhancement of Power Systems

In this chapter, the hardware and software design and implementation have been carried out for the discussed wide-area damping control (WADC). First, a hardware-in-the-loop (HIL) test system based on the RT-LAB platform® has been established. Three WADC algorithms, i.e., the phase-compensation method, the proposed delay-dependent state-feedback method [1], and the proposed delay-dependent dynamic output-feedback method [2], are introduced. Then, the software design of the above control algorithms has been presented, and the bilinear transform method is employed to get the discrete-time model of the controllers. The designed WADC is embedded in a real-time power system. Finally, an experimental study is carried out based on the HIL simulation. A typical interconnected power system is modeled in RT-LAB®, the closed-loop test has been done to validate the proposed control concept and controller design methods even in conditions with time-varying delays, and compare the damping performance of the controller using different control algorithms.

11.1 System Description

In order to test the performance of the designed hardware controller for the wide-area damping strategies, a hardware-in-the-loop (HIL) real-time simulation has been carried out with an IEEE benchmark system [3, 4].

In this study, the IEEE benchmark system is modified by placing one shunt-type FACTS (SVC) device at the middle of the interconnected line.

Figure 11.1 shows the real-time simulation and the interface of the simulation model with the hardware controller. Two phasor measurement units (PMUs) are used to monitor the rotor speed of G1 and G2. The WAMS center is in charge of receiving the wide-area signal from PMU. The hardware controller is regarded as the functional extension of the WAMS. It receives the control-input signal from the WAMS center, and sends a control-output signal to the FACTS device while the control algorithm is computed. The combination of WAMS center and hardware controller forms an enhanced WAMS center with the wide-area stability control. Besides, it should be noted that the control-output signal is in fact a supplementary

© Springer-Verlag Berlin Heidelberg 2016
Y. Li et al., *Interconnected Power Systems*, Power Systems,
DOI 10.1007/978-3-662-48627-6_11

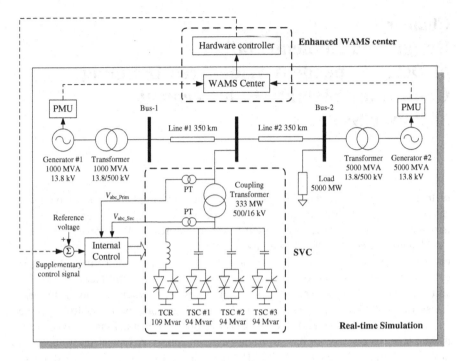

Fig. 11.1 Real-time simulation model with hardware-in-the-loop (*HIL*)

control signal of the internal control of the FACTS device. In this way, the FACTS controller has two control functions: the bus-voltage stability control (local control) and the oscillation damping control (wide-area control).

The modal analysis indicates that the original system has a dominant oscillatory mode (frequency 0.978 Hz, damping ratio −0.0159). Here to illustrate the performance of the designed hardware controller on damping inter-area oscillatory mode, the inertia coefficient of G1 and G2 is modified from 3.7 to 7.4 s. In this way, the modified system has a typical inter-area oscillatory model (frequency 0.674 Hz, damping ratio 0.00619). Moreover, Fig. 11.2 shows the established HIL test system based on the RT-LAB platform®.

11.2   Hardware Design

Figure 11.3 shows the hardware configuration of the wide-area damping controller. The hardware controller mainly contains four parts: the ADC boosted circuit, the microcontroller, the DAC and conditioning circuit and the power source. The function of each part is summarized as follows:

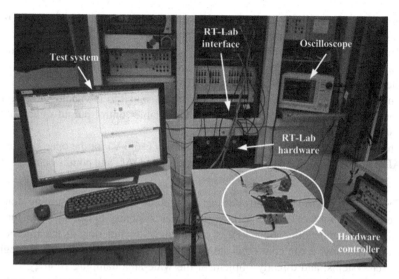

Fig. 11.2 Setup of the HIL test system based on the RT-LAB platform®

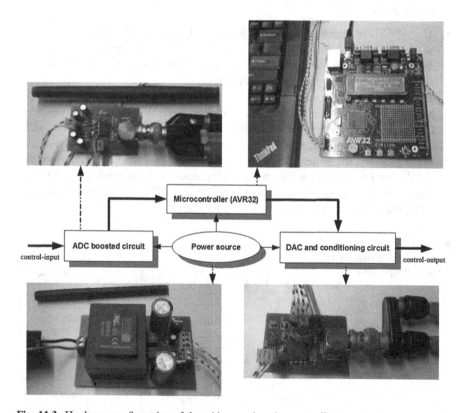

Fig. 11.3 Hardware configuration of the wide-area damping controller

- Since the ADC can only identify the positive voltage signal, the boosted circuit has to be designed to increase the control-input signal more than 0 V.
- The microcontroller receives the analog signal from the ADC boosted circuit, implements the control algorithm of the wide-area damping strategy, and sends the digital control-output signal to the DAC.
- The DAC is in charge of converting the digital signal to the analog signal. The conditioning circuit is also included to get a suitable control-output signal for the controllable device in power systems.
- The power source provides the dc power supply (± 6.0 V) for the operation of the ADC boosted circuit, the microcontroller, and the DAC and conditioning circuit.

The proposed hardware controller can be used as a wide-area damping control unit embedded in the WAMS center of power systems. It can also be further developed by embedding the wireless communication function in the microcontroller. In this way, it can communicate with the PMU located in the remote region of power systems, and directly implement the wide-area damping strategy.

Figure 11.4 shows the flowchart of the hardware implementation for these WADC strategies presented in this dissertation. The flowchart mainly includes the following parts:

- The ADC is in charge of sampling the control-input signal measured from the WAMS. The sampling time Ts is controlled by the time counter (TC). Here, Ts = 100 μs, which means that in each cycle (1/50(Hz) = 0.02 s) there are 200 signals transmitted to the microcontroller.
- The direct memory access (DMA) channel is in charge of the communication between the memory and the peripheral device. Here, two channels are selected. DMA1 and DMA2 are used to transmit the control-input signal from the ADC or the control-output signal from the memory to the DAC, respectively.
- In the memory, the buffer adcRX is defined to store the sampled signal from the ADC. The central processing unit (CPU) processes the control algorithm and stores the control-output signal to the buffer algo. Then, under the control of CPU, the signal is transmitted to the buffer dacTX.

Fig. 11.4 Flowchart of the hardware implementation of the control algorithm

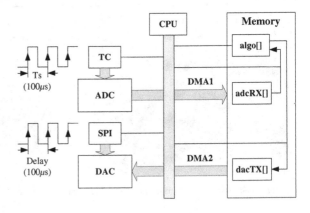

- The DAC is used to process the control-output signal from dacTX. The digital signal is converted to an analog signal. Like the ADC, a serial peripheral interface is used to control the sampling time of signal transmission. Here, the delay is set to 100 μs, which means that in each cycle (0.02 s) the DAC finishes to transmit 200 control-output signals.

Moreover, Fig. 11.5 shows the flowchart of the interrupt program for the hardware implementation. In this flowchart, the "adc_dma_full" evaluate which buffer (RX1 or RX2) should be selected to store the sampled signal from ADC; the "algo_TX_spi"

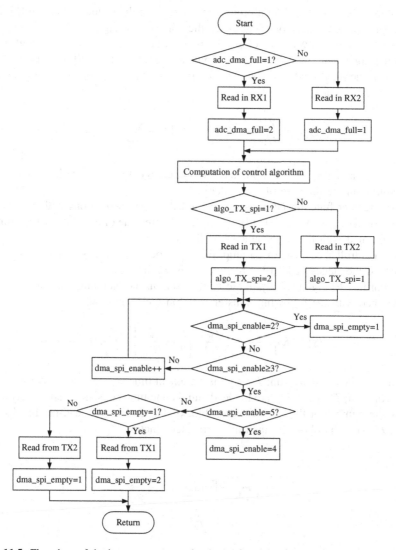

Fig. 11.5 Flowchart of the interrupt program for the hardware implementation

decides if buffer TX1 or TX2 should be selected to store the control-output signal from the computation result of the control algorithm; and the "dma_spi_empty" assesses the DMA2 to read the control-output signal from the buffer TX1 or TX2.

11.3 Design of the Control Algorithm

11.3.1 Classic Phase Compensation Method

The damping controller design procedure is based on the residue approach [5]. For the approach, the values of residues for the controller design can be calculated.

Figure 11.6 shows the closed-loop system with a damping controller. Considering the system with single input and single output (SISO), the residue matrix of the plant model can be obtained from the transfer-function representation of the system that is:

$$G(s) = \frac{Y(s)}{U(s)} = C(sI - A)^{-1}B = \sum_{i=1}^{n} \frac{R_i}{(s - \lambda_i)} \quad (11.1)$$

where $U(s)$ and $Y(s)$ are the input and the output signal of the linearized system, $G(s)$ is the transfer function of the system.

The transfer function $G(s)$ can be expanded in partial fractions with residue R_i and eigenvalue λ_i that is (11.1) [5], where R_i is the residue of the transfer function $G(s)$ associated with the eigenvalues λ_i at the mode $-i$.

The controller consists of an amplification block, a high-pass filter, a low-pass filter, and mc stages of lead-lag blocks, as shown in Fig. 11.7.

According to Fig. 11.7, by phase compensation, the damping controller can be determined with the following transfer function representation:

$$H(s) = K \frac{1}{1 + sT_m} \frac{sT_w}{1 + sT_w} \left[\frac{1 + sT_{\text{lead}}}{1 + sT_{\text{lag}}} \right]^{m_c} = KH_1(s) \quad (11.2)$$

where K is the constant gain; T_m is a measurement time constant; T_w is the washout time constant; T_{lead} and T_{lag} are the lead and the lag time constant, respectively, and m_c is the number of the compensation blocks. Generally, two lead-lag blocks are used. $H_1(s)$ is the phase compensation transfer function.

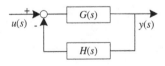

Fig. 11.6 Closed-loop system with a damping controller

Control
input

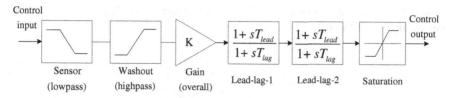

Sensor   Washout   Gain
(lowpass)   (highpass)   (overall)   Lead-lag-1   Lead-lag-2   Saturation

Fig. 11.7 Structure of a damping controller

For the closed-loop system with the feedback transfer function $H(s)$, the eigenvalue sensitivity is shown in (11.3), which gives the relationship between the sensitivity of eigenvalue to feedback loop gain and the open-loop residue associated with the same eigenvalue.

$$\frac{\partial \lambda_i}{\partial K} = R_i \frac{\partial H(\lambda_i)}{\partial K} = R_i \frac{\partial K H_1(\lambda_i)}{\partial K} = R_i H(\lambda_i) \tag{11.3}$$

Therefore assuming the gain K is small enough and adding the feedback to the system, it will cause a change in eigenvalue in the initial operating point [6].

$$\Delta \lambda_i = R_i K H(\lambda_i) \tag{11.4}$$

From (11.3), it can be observed that when the feedback control $H(s)$ is applied, the eigenvalues of the initial system will be changed. The objective of the FACTS damping controller is to improve the damping ratio of the selected oscillation model i, and $\Delta \lambda_i$ must be a real negative value to move the real part of the eigenvalue to the left half negative complex plane without changing the frequency.

Figure 11.8 shows the movement of the eigenvalue of the system with a FACTS-based WADC controller. The compensation angle is used to move the eigenvalue direct to the negative axis. In this figure, the phase angle φ_{com} shows the compensation angle, which is needed to move the eigenvalue directly to the left parallel with the real axis. The phase angle φ_{com} can be achieved via the lead-lag

Fig. 11.8 Concept of phase compensation

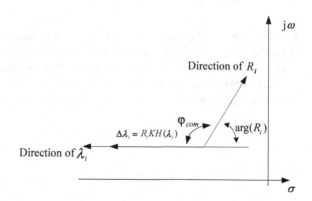

function. The parameters of the lead and lag time constants, T_{lead} and T_{lag}, can be determined as follows [6]:

$$\varphi_{\text{com}} = 180° - \arg(R_i)$$

$$\alpha = \frac{T_{\text{lead}}}{T_{\text{lag}}} = \frac{1 - \sin\frac{\varphi_{\text{com}}}{m_c}}{1 + \sin\frac{\varphi_{\text{com}}}{m_c}}$$

$$T_{\text{lag}} = \frac{1}{\omega_n\sqrt{\alpha}} \tag{11.5}$$

$$T_{\text{lead}} = \alpha T_{\text{lag}}$$

where φ_{com} is the phase angle; $\arg(R_i)$ is the phase angle of the residue R_i, ω_n is the frequency of the dominant oscillation mode. The parameter m_c is the number of the lead-lag blocks. Usually two blocks are sufficient to compensate the input signal, while the angle compensated by each block should not exceed 60° [6].

The WADC controller gain K has to be optimized for the damping of all modes of oscillation because it is proportional for the feedback of the closed-loop system and the controller gain. The damping effectiveness of the WADC controller is influenced by K. As shown in Fig. 11.8, the eigenvalue λ_i from the original operating point with the increase of K will move to the negative axis. Therefore, the K can be computed as a function of the desired eigenvalue location according to the following equation:

$$\Delta\lambda_i = \left|\frac{\Delta\lambda_i}{R_iH_1(\lambda_i)}\right| = \left|\frac{\lambda_{i,\text{des}} - \lambda_i}{R_iH_1(\lambda_i)}\right| \tag{11.6}$$

where $\lambda_{i,\text{des}}$ is the desired eigenvalue. The damping of i-mode oscillation can be enhanced once a greater K has been chosen. Therefore, the value of K can be determined using the root-locus method [7].

11.3.2   Delay-Dependent State-Feedback Robust Design Method

For the wide-area damping controller design, the objective is to solve the following delay-dependent robust H_∞ control problem: with a given H_∞ performance index γ, design the H_∞ controller (11.7) to ensure that the closed-loop system (11.8) is internal stability and at the zero initial condition, the closed-loop system (11.8) holds the H_∞ performance index γ under external disturbances that is

$$\|z(t)\|_\infty < \gamma\|w(t)\|_\infty \tag{11.7}$$

The memory-less state-feedback controller is

$$u(t) = Kx(t - \tau(t)) \tag{11.8}$$

where K is the control gain.

The closed-loop system, which considers the time delay, can be obtained as

$$T(s) = \begin{cases} \dot{x}(t) = Ax(t) + BKx(t - \tau(t)) \\ z(t) = \begin{bmatrix} Cx(t) + D_w w(t) \\ D_u Kx(t - \tau(t)) \end{bmatrix} \\ y(t) = Cx(t) + D_w w(t) \\ x(t) = 0,\ t \in [-h\,0] \end{cases} \tag{11.9}$$

where $x(t)$, $u(t)$, $z(t)$, and $w(t)$ are the system state, control input, disturbance output, and disturbance input, respectively, A, B, C, D_{w}, and D_u are the state matrix, control-input matrix, system output matrix, system disturbance output matrix, and control disturbance output matrix, respectively; h is the maximum value of the varying delay $\tau(t)$ which satisfies

$$0 \le \tau(t) \le h,\ \dot{\tau}(t) \le d < 1,\ \forall t \ge 0 \tag{11.10}$$

Theorem 1 [8] *Considering the closed-loop system* (11.9) *with the scalars h and d as given in* (11.10), *and the H_∞ performance index γ as given in* (11.7), *if there exist matrices $P > 0$, $R > 0$, $Q > 0$, free-weighting matrices M_1, M_2, and the positive value ε, such that the following LMI holds:*

$$\Xi = \begin{bmatrix} \phi_{11} & \phi_{12} & PB_w & hA_{0K}^T & hM_1^T & C & K^T D_u^T \\ * & \phi_{22} & 0 & hA_1^T & hM_2^T & 0 & 0 \\ * & * & -\gamma^2 I & hB_w^T & 0 & D_w^T & 0 \\ * & * & * & -hR^{-1} & 0 & 0 & 0 \\ * & * & * & * & -hR & 0 & 0 \\ * & * & * & * & * & -I & 0 \\ * & * & * & * & * & * & -I \end{bmatrix} < 0 \tag{11.11}$$

where:

$$\phi_{11} = PA + A^T P + Q + M_1^T + M_1,$$
$$\phi_{12} = PBK - M_1^T + M_2,$$
$$\phi_{22} = -(1-d)Q - M_2^T + M_2$$

then the closed-loop system (11.9) maintains the internal stability and the external H_∞ performance index γ.

Theorem 1 can be further converted into a standard generalized nonlinear optimization problem by means of the congruent transformation. Define the following matrices:

$$\prod = \text{diag}\{P^{-1},\, P^{-1},\, I,\, I,\, P^{-1},\, I,\, I\} \tag{11.12}$$

Pre- and post-multiply Ξ in (11.11) by $\prod$ in Eq.(11.12), and introduce a new matrix S, defined as $P^{-1}RP^{-1} \geq S$, to deal with the nonlinear term, then the following equivalent LMIs can be obtained:

$$\begin{bmatrix} v_{11} & v_{12} & B_w & v_{13} & hN_1^T & P^{-1}C^T & Y^TD_u^T \\ * & v_{22} & 0 & hP^{-1}A^T & hN_2^T & 0 & 0 \\ * & * & -\gamma^2 I & hB_w^T & 0 & D_w^T & 0 \\ * & * & * & -hR^{-1} & 0 & 0 & 0 \\ * & * & * & * & -hS & 0 & 0 \\ * & * & * & * & * & -I & 0 \\ * & * & * & * & * & * & -I \end{bmatrix} < 0 \tag{11.13}$$

$$\begin{bmatrix} S^{-1} & L^{-1} \\ L^{-1} & \bar{R}^{-1} \end{bmatrix} \geq 0 \tag{11.14}$$

where:

$$v_{11} = AP^{-1} + P^{-1}A^T + BY + Y^TB^T + P^{-1}QP^{-1} + N_1^T + N_1,$$
$$v_{12} = BKP^{-1} - N_1^T + N_2,\ v_{13} = hP^{-1}A^T + hY^TB^T,$$
$$v_{22} = -(1-d)P^{-1}QP^{-1} - N_2^T - N_2,\ N_1 = P^{-1}M_1P^{-1},$$
$$N_2 = P^{-1}M_2P^{-1},\ Y = KP^{-1}.$$

If there exist real matrices $P^{-1} > 0,\, R^{-1} > 0,\, P^{-1}QP^{-1} > 0,\, N_1,\, N_2$, and Y, so that (11.13) and .(11.14) can be held, then the closed-loop system (11.9) can reach the internal stability and hold the H_∞ performance index γ. In addition, the controller (11.8) can be obtained as $K = YL^{-1}$

$$K(s) = \begin{cases} x_c(t-\tau(t)) = A_c x_c(t-\tau(t)) + B_c u_2(t-\tau(t)) \\ y_2 = C_c x_c(t-\tau(t)) + D_c y(t-\tau(t)) \end{cases} \tag{11.15}$$

Further, according to the concept of cone complementarity problem in [12], the calculation of the HVDC-WASC (11.15) [9] can be converted as the calculation of a nonlinear minimization problem expressed as

$$\text{Min. Trace}\left(SS^{-1} + S^{-1}S + P^{-1}P + PP^{-1} + R^{-1}R + RR^{-1}\right) \tag{11.16}$$

Subject to (11.13) and

$$
\begin{cases}
\begin{bmatrix} S^{-1} & P \\ P & R \end{bmatrix} \geq 0, \ P^{-1} > 0, \ R > 0, \ Q > 0, \\[4mm]
\begin{bmatrix} S & I \\ I & S^{-1} \end{bmatrix} \geq 0, \ \begin{bmatrix} P^{-1} & I \\ I & P \end{bmatrix} \geq 0, \ \begin{bmatrix} R^{-1} & I \\ I & R \end{bmatrix} \geq 0.
\end{cases}
\tag{11.17}
$$

11.3.3 Delay-Dependent Dynamic Output-Feedback Control Method

Regarding an interconnected system, its linearized discrete-time model considering signal time-varying delay can be formulated as:

$$
\begin{cases}
x(k+1) = Ax(k) + A_d x(k - d(k)) + Bu(k) \\
y(k) = Cx(k) + C_d x(k - d(k)) \\
x(k) = \phi(k), \ -d_2 \leq k \leq 0
\end{cases}
\tag{11.18}
$$

where $x(k)$ is the state vector; $u(k)$ and $y(k)$ are wide-area control-output and control-input; A, A_d, B, C and C_d are constant matrices and $d(k)$ is the time-varying delay that satisfies the following condition:

$$
d_1 \leq d(k) \leq d_2
\tag{11.19}
$$

The objective of the controller design is to work out a dynamic control gain represented by (11.18), so that the closed-loop system can reach asymptotically stable when the controller suffers a certain time-varying delay.

$$
\begin{cases}
x_c(k+1) = A_c x_c(k) + B_c y(k) \\
u(k) = C_c x_c(k) + D_c y(k) \\
x_c(k) = 0, \ k < 0
\end{cases}
\tag{11.20}
$$

where $x_c(k)$ is the state vector of the dynamic control gain and A_c, B_c, C_c, and D_c are constant matrices to be optimized.

For facilitating controller design in the LMI framework, the dynamic control gain (11.20) is transformed to the following representation:

$$
u(k) = \tilde{K}_{\text{DOF}} y(k) = \begin{bmatrix} D_c & C_c \\ B_c & A_c \end{bmatrix} y(k)
\tag{11.21}
$$

Then, under the action of (11.19), the open-loop system (11.20) can yield a closed-loop system with the following representation:

$$\begin{cases} \xi(k+1) = \tilde{A}\xi(k) + \tilde{A}_d\xi(k-d(k)) + \tilde{B}u(k) \\ y(k) = \tilde{C}\xi(k) + \tilde{C}_d\xi(k-d(k)) \end{cases} \qquad (11.22)$$

where $\quad \xi(k) = \begin{bmatrix} x^T(k) \\ x_c^T(k) \end{bmatrix}, \quad \tilde{A} = \begin{bmatrix} A & 0 \\ 0 & 0 \end{bmatrix}, \quad \tilde{A}_d = \begin{bmatrix} A_d & 0 \\ 0 & 0 \end{bmatrix}, \quad \tilde{B} = \begin{bmatrix} B & 0 \\ 0 & I \end{bmatrix},$

$\tilde{C} = \begin{bmatrix} C & 0 \\ 0 & I \end{bmatrix}, \tilde{C}_d = \begin{bmatrix} C_d & 0 \\ 0 & 0 \end{bmatrix}.$

In this way, the optimization of (11.20) is transformed to that of (11.21). The stability criterion of (11.18) is analyzed for the DOF-WADC design. Nowadays, there are various stability analysis methods formulated in the LMI framework [10]. However, these methods cannot sufficiently deal with the effect of time-varying delay on stability performance. A new theorem normalized in recent years is applied to analyze the stability criterion taking time-varying delay into consideration.

Theorem 2 [11] *For given scalars d_1 and d_2 with $d_2 > d_1 > 0$, the system* (11.18) *with $d(k)$ satisfying* (11.19) *can be stabilized by* (11.20) *if there exist $P = P^T > 0$, $L = L^T > 0$, $Q_i = Q_i^T \geq 0$ (i = 1, 2, 3), $Z_j = Z_j^T > 0$, $R_j = R_j^T > 0$ (j = 1, 2),* $X = \begin{bmatrix} X_{11} & X_{12} \\ * & X_{22} \end{bmatrix} \geq 0$, $Y = \begin{bmatrix} Y_{11} & Y_{12} \\ * & Y_{22} \end{bmatrix} \geq 0$, $N = [N_1^T \ N_2^T]^T$, $S = [S_1^T \ S_2^T]^T$, $M = [M_1^T$ $M_2^T]^T$, and $\tilde{K}_{DOF}$ satisfies that,

$$\psi_1 = \begin{bmatrix} X & N \\ * & Z_1 \end{bmatrix} \geq 0 \qquad (11.23)$$

$$\psi_2 = \begin{bmatrix} Y & S \\ * & Z_2 \end{bmatrix} \geq 0 \qquad (11.24)$$

$$\psi_3 = \begin{bmatrix} X+Y & M \\ * & Z_1+Z_2 \end{bmatrix} \geq 0 \qquad (11.25)$$

$$PL = I, Z_jR_j = I \ (j = 1, 2) \qquad (11.26)$$

$$\begin{bmatrix} \phi & \Xi_1^T & d_2\Xi_2^T & d_{12}\Xi_2^T \\ * & -L & 0 & 0 \\ * & * & -d_2R_1 & 0 \\ * & * & * & -d_{12}R_2 \end{bmatrix} < 0 \qquad (11.27)$$

where

$$
\phi = \begin{bmatrix}
\phi_{11} & \phi_{12} & S_1 & -M_1 \\
* & \phi_{22} & S_2 & -M_2 \\
* & * & -Q_1 & 0 \\
* & * & * & -Q_2
\end{bmatrix}
$$

$\phi_{11} = Q_1 + Q_2 + (d_{12}+1)Q_3 - P + N_1 + N_1^T + d_2 X_{11} + d_{12} Y_{11},$

$\phi_{12} = N_2^T - N_1 + M_1 - S_1 + d_2 X_{12} + d_{12} Y_{12}$

$\phi_{22} = -Q_3 - N_2 - N_2^T + M_2 + M_2^T - S_2 - S_2^T + d_2 X_{22} + d_{12} Y_{22},$

$\Xi_1 = \begin{bmatrix} \tilde{A} + \tilde{B}\tilde{K}_{DOF}\tilde{C} & \tilde{A}_d + \tilde{B}\tilde{K}_{DOF}\tilde{C}_d & 0 & 0 \end{bmatrix}$

$\Xi_2 = \begin{bmatrix} \tilde{A} + \tilde{B}\tilde{K}_{DOF}\tilde{C} - I & \tilde{A}_d + \tilde{B}\tilde{K}_{DOF}\tilde{C}_d & 0 & 0 \end{bmatrix}$

Since some equality terms like (11.26) are existed in the theorem, it is impossible to optimize (11.21). To overcome this problem, a cone complementary linearization (CCL) [12] is used to transform the controller design to the following standard optimization problem:

$$
\text{Min.Trace}\left(PL + \sum_{j=1}^{2} Z_j R_j\right) \tag{11.28}
$$

Subject to (11.23)–(11.25), (11.27), and

$$
\begin{bmatrix} P & I \\ I & L \end{bmatrix} \geq 0, \quad \begin{bmatrix} Z_j & I \\ I & R_j \end{bmatrix} \geq 0, \, j = 1, 2 \tag{11.29}
$$

$$
P > 0, \, L > 0, \, Q_i > 0, \, i = 1, 2, 3 \tag{11.30}
$$

$$
Z_j > 0, \, R_j > 0, \, j = 1, 2 \tag{11.31}
$$

11.4 Algorithm Implementation

11.4.1 Discrete-Time Model for the Hardware Controller

Generally, the controller designed by the robust method represents a continuous-time form. However, in practice, it is impossible to directly apply such form to the hardware implementation, thus a suitable discrete method has to be used to transform it into a discrete-time form.

In this section, the bilinear transformation [13], also called Tustin's method, is selected for the discrete modeling of the wide-area damping controller. By the bilinear transformation, the relationship between s-plane and z-plane can be obtained as:

$$s = \frac{2\,(z-1)}{T\,(z+1)} \tag{11.32}$$

where T is the sampling time.

In the s-plane, s is made up of the real-part σ and the imaginary-part ω that is $s = \sigma + j\omega$. The above (11.32) can be rewritten as the following amplitude form:

$$|z| = \left|\frac{1 + \frac{T}{2}s}{1 - \frac{T}{2}s}\right| = \frac{\sqrt{\left(1 + \frac{T}{2}\sigma\right)^2 + \left(\frac{\omega T}{2}\right)^2}}{\sqrt{\left(1 + \frac{T}{2}\sigma\right)^2 - \left(\frac{\omega T}{2}\right)^2}} \tag{11.33}$$

According to (11.33), it can be seen that: (1) if σ in the s-plane is equal to 0 that is $s = j\omega$, then $|z|$ will be equal to 1 ($|z| = 1$) that is a unit circle in the z-plane; (2) if $\sigma > 0$, then $|z| > 1$; (3) if $\sigma < 0$, then $|z| < 1$. In this way, the relation mapping between the s- and the z-plane can be obtained, as shown in Fig. 11.9.

Three types of control algorithm for the wide-area damping strategy have been proposed, i.e., the classic phase compensation method, a delay-dependent robust design method and the dynamic output-feedback control method.

In order to evaluate these control methods, a real-time simulation model of an IEEE benchmark test system is established in the environment of RT-LAB®. The hardware controller, designed for wide-area damping strategy, will be applied to this test system to compare the damping performance of the controller using different control methods.

According to the control principle presented in the above sections, different types of wide-area damping controller can be designed for the test system. More specifically, Table 11.1 shows the continuous-time transfer function $H(s)$ of each controller. It can be seen that $H(s)$ contains different representations of the transfer function that is

Fig. 11.9 Relation mapping of the bilinear transformer in the s- and the z-plane

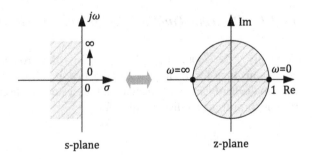

Table 11.1 Transfer functions of different types of wide-area damping controller

Type	$H(s)$
Phase compensation (PC-WADC)	$\frac{1}{1+0.015s}\frac{0.7s}{1+0.7s}20.4256\frac{1+0.111s}{1+s}\frac{1+21.5879s}{1+s}$
Delay-dependent robust control (DD-WADC)	$\frac{10s}{10s+1}\frac{939.4s+2276}{s^2+36.14s+74.33}$
Dynamic output-feedback control (DOF-WADC)	$\frac{10s}{10s+1}30\frac{1.807s+8.513}{s+4.796}$

1. The low-pass filter (LPF) expressed by:

$$H(s)_{\text{LPF}} = \frac{1}{1+T_{\text{LPF}}s} \tag{11.34}$$

where T_{LPF} is the time constant of the LPF.
According to (11.32), the discrete-time form of (11.34) can be represented as:

$$H(z)_{\text{LPF}} = H(s)_{\text{LPF}}\Big|_{s=\frac{2}{T}\frac{z-1}{z+1}} = \frac{\frac{T}{T+2T_{\text{LPF}}}z + \frac{T}{T+2T_{\text{LPF}}}}{z + \frac{T-2T_{\text{LPF}}}{T+2T_{\text{LPF}}}} \tag{11.35}$$

2. The high-pass filter (HPF) expressed by:

$$H(s)_{\text{HPF}} = \frac{T_{\text{HPF}}s}{1+T_{\text{HPF}}s} \tag{11.36}$$

where T_{HPF} is the time constant of the HPF.
According to (11.32), the discrete-time form of (11.36) can be represented as:

$$H(z)_{\text{HPF}} = H(s)_{\text{HPF}}\Big|_{s=\frac{2}{T}\frac{z-1}{z+1}} = \frac{\frac{2T_{\text{HPF}}}{T+2T_{\text{HPF}}}z - \frac{2T}{T+2T_{\text{HPF}}}}{z + \frac{T-2T_{\text{HPF}}}{T+2T_{\text{HPF}}}} \tag{11.37}$$

3. The lead-lag block $H(s)_{\text{LL}}$ in the PC-WADC or the dynamic gain $H(s)_{\text{DOF}}$ in the DOF-WADC. Both of them have the following similar representation:

$$H(s)_{\text{LL or DOF}} = \frac{b_1 s + b_0}{a_1 s + a_0} \tag{11.38}$$

where b_1 and b_0 are the numerator coefficients, in the $H(s)_{\text{LL}}$, $b_0 = 1$; a_1 and a_0 are the denominator coefficients, in the $H(s)_{\text{LL}}$, $a_0 = 1$, and in the $H(s)_{\text{DOF}}$, $a_1 = 1$.

According to (11.32), the discrete-time form of (11.38) can be represented as:

$$H(z)_{\text{LL or DOF}} = H(s)_{\text{LL or DOF}}\Big|_{s=\frac{2}{T}\frac{z-1}{z+1}} = \frac{\frac{2b_1+b_0T}{2a_1+a_0T}z + \frac{-2b_1+b_0T}{2a_1+a_0T}}{z + \frac{-2a_1+a_0T}{2a_1+a_0T}} \tag{11.39}$$

The transfer function $H(s)_{DD}$ in the DD-WADC is constructed by the state observer and the control gain. It has the following representation:

$$H(s)_{DD} = \frac{b_1 s + b_0}{a_2 s^2 + a_1 s + a_0} \tag{11.40}$$

where b_1 and b_0 are the numerator coefficients; a_i ($i = 0, 1, 2$) is the denominator coefficient, and $a_2 = 1$.

According to (11.32), the discrete-time form of (11.40) can be represented as:

$$H(z)_{DD} = H(s)_{DD}\big|_{s=\frac{2z-1}{Tz+1}}$$

$$= \frac{\frac{2b_1 T + b_0 T^2}{4a_2 + 2a_1 T + a_0 T^2} z^2 + \frac{2b_0 T^2}{4a_2 + 2a_1 T + a_0 T^2} z + \frac{-2b_1 T + b_0 T^2}{4a_2 + 2a_1 T + a_0 T^2}}{z^2 + \frac{-8a_2 + 2a_0 T^2}{4a_2 + 2a_1 T + a_0 T^2} z + \frac{4a_2 - 2a_1 T + a_0 T^2}{4a_2 + 2a_1 T + a_0 T^2}} \tag{11.41}$$

To verify the effectiveness of the discrete-time models obtained by the bilinear transform, Fig. 11.10 gives the bode diagrams of the LPF, the HPF, the dynamic gain in the DOF-WADC, and the transfer function in the DD-WADC. From this figure, it can be seen that the frequency response of each discrete-time model coincides with that of each continuous-time model very well. This confirms the discretized results on these wide-area damping controllers.

11.4.2   Algorithm Flowchart

According to the above subsection, it can be seen that the discrete-time models of the proposed WADC contain two general expressions

$$H(z)_1 = \frac{Y_1(z)}{U_1(z)} = \frac{B_1 z + B_0}{z + A_0} \tag{11.42}$$

$$H(z)_2 = \frac{Y_2(z)}{U_2(z)} = \frac{B_1 z^2 + B_1 z + B_0}{z^2 + A_1 z + A_0} \tag{11.43}$$

where $Y_1(z)$ and $Y_2(z)$ are the output signal of $H(z)_1$ and $H(z)_2$; $U_1(z)$ and $U_2(z)$ are the input signals of $H(z)_1$ and $H(z)_2$.

The recursion expressions of (11.42) and (11.43) can be determined as follows:

$$y_1(k) = B_1 x_1(k) + B_0 x_1(k-1) - A_0 y_1(k-1) \tag{11.44}$$

$$y_2(k) = B_2 x_2(k) + B_1 x_2(k-1) + B_0 x_2(k-2) \\ - A_1 y_2(k-1) - A_0 y_2(k-2) \tag{11.45}$$

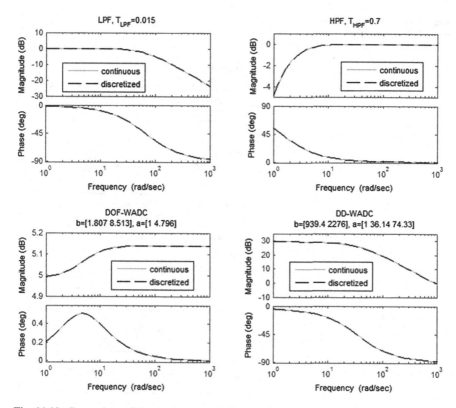

Fig. 11.10 Comparison of the continuous and discretized responses

where k is the discrete time; $y_1(k)$ and $y_2(k)$ are the output signals in the time k; $x_1(k)$ and $x_2(k)$ are the input signals in the time k; $y_1(k-1)$ and $y_2(k-1)$ are the output signals in the time $k-1$; $x_1(k-1)$ and $x_2(k-1)$ are the input signals in the time $k-1$; $y_2(k-2)$ and $x_2(k-2)$ are the output and the input signals in the time $k-2$.

Figures 11.11, 11.12 and 11.13 shows the algorithm flowcharts of the discussed wide-area damping strategies, i.e., the conventional PC-WADC, the new DD-WADC and the new DOF-WADC. It can be seen from Fig. 11.11 that the conventional PC-WADC contains four discrete-time models that is one HPF, one LPF and two lead-lag blocks. In comparison, the new DD-WADC and DOF-WADC have much simpler structure. Both of them only have one HPF and one control transfer function. Moreover, the order of the control transfer function of the new DOF-WADC is 2, while that of the new DD-WADC is 1. So the DD-WADC is the simplest among these three controllers.

Fig. 11.11 Flowchart of the control algorithm of the PC-WADC

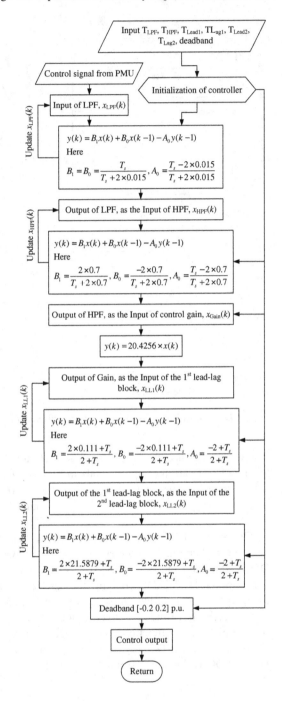

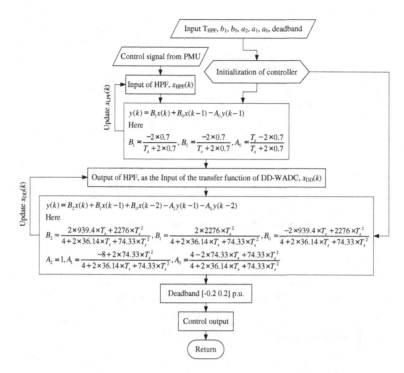

Fig. 11.12 Flowchart of the control algorithm of the DD-WADC

11.5   Experimental Results

An experimental study is carried out in order to validate the discussed wide-area damping control strategies. And the hardware including the microcontroller, the peripheral analog to digital converter and digital to analog converter has been implemented based on the AVR 32bit development board [14]. The interrupt flowchart and the main program are also proposed for the hardware controller. By means of the RT-LAB platform® [15], a HIL simulation is carried out to test the designed hardware controller under consideration of signal time-varying delays.

A line-to-ground fault (big disturbance) near Bus-1 is used to test the performance of the hardware controller in damping inter-area oscillation of the interconnected system.

Figure 11.14 shows the experimental result about the response of the speed difference dw12 between Generator #1 and #2.

From the results shown in Fig. 11.14a it can be seen that when the system is open-loop without the hardware controller, the big disturbance excites a serious inter-area oscillation. Moreover, the time-varying delay τ of the wide-area communication network is also considered, so the signal measured at the control center (the enhanced WAMS center) delays this measured from the PMU. When the

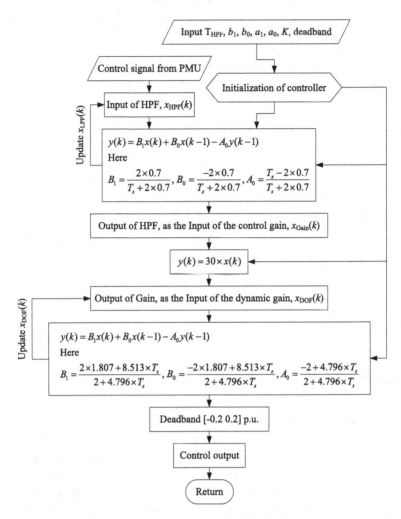

Fig. 11.13 Flowchart of the control algorithm of the DOF-WADC

hardware controller adopts the traditional PC-WADC, as shown in Fig. 11.14b, although it can damp the oscillation to a certain degree, the damping performance is easily affected by the time-varying delay of the communication network. When the delay τ changes from 800 to 250 ms, a serious oscillation is excited. In such a case, the PC-WADC even provides the negative damping and the system is facing instability. From Fig. 11.14c, d we can see that the DD-WADC and the DOF-WADC proposed in this dissertation can effectively damp inter-area oscillation even under consideration of time-varying delays. Both of them represent a good robustness on the delay of the communication network. Besides, by comparing Fig. 11.14c, d, we can further see that the DOF-WADC has more effective damping than the DD-WADC. As the above mentioned, comparing with the

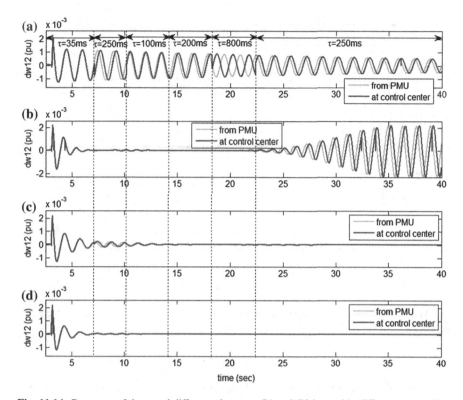

Fig. 11.14 Response of the speed difference between G1 and G2 located in different areas of the interconnected system, (**a**) no WADC; (**b**) PC-WADC; (**c**) DD-WADC; (**d**) DOF-WADC

structure of the PC-WADC and the DD-WADC, the DOF-WADC is the simplest, which could easily be applied in the hardware.

Figure 11.15 shows the control-output responses of the hardware controller using different control algorithms. As shown in Fig. 11.12, the WADC is the supplementary control of the internal controller (local voltage stability control, the reference voltage V_{ref} is 1.0 p.u.). To guarantee this local control performance, the control-output of the WADC is set in $[-0.2\ 0.2]$ p.u. From Fig. 11.15 it can be seen that due to the delay varying from 800 to 250 ms, although the PC-WADC reaches the limitation of the control-output, it still cannot damp the oscillation effectively, so that the control-output oscillates in the range of $[-0.2\ 0.2]$ p.u. Unlike this, it can also be seen from Fig. 11.15 that the DD- and the DOF-WADC can provide the effective control-output signal whatever the varying delays of the communication network.

Furthermore, to illustrate the performance of the designed hardware controller on the stability of the interconnected system, Fig. 11.16 shows the response of the power flow through the interconnected line (Line 1# in Fig. 11.1). From Fig. 11.16a it can be seen that the power oscillations exist in the open-loop system without

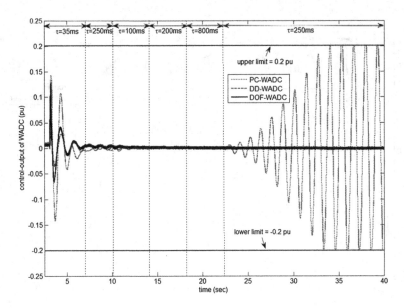

Fig. 11.15 Response of the control-output of the hardware controller using different control algorithms

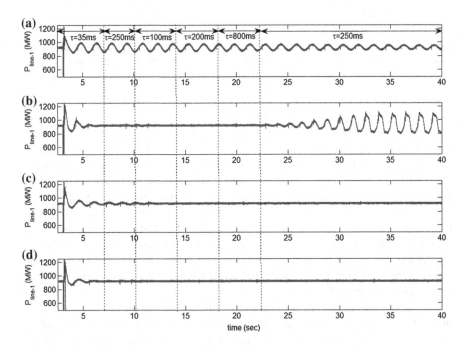

Fig. 11.16 Response of the power flow through the interconnected line of the test system, (a) no WADC; (b) PC-WADC; (c) DD-WADC; (d) DOF-WADC

hardware controller. The traditional PC-WADC can damp the power oscillation, but it cannot maintain the damping performance under the effect of the time-varying delay, as shown in Fig. 11.16b. Both the DD- and DOF-WADC can reach good damping performance even in conditions with time-varying delays. The performance of the DOF-WADC is better than that of the DD-WADC.

11.6 Summary

In this chapter, the hardware implementation and the experiment have been carried out for the wide-area damping control strategies. The bilinear transform method is employed to get the discrete-time model of the controller. Based on this, the flowchart has been proposed for implementing the control algorithms in the environment of hardware. Then, the hardware structure of the wide-area damping controller has been designed, which includes the ADC boosted circuit, the microcontroller, the DAC and conditioning circuit, and the power source. The interrupt program of the hardware implementation is also proposed. Finally, the HIL real-time simulation based on the RT-LAB platform® has been done to demonstrate the performance of the hardware controller.

The experimental results indicate that the hardware using the DD- or the DOF-WADC algorithms can effectively damp inter-area oscillation of the interconnected system. Even under the effect of time-varying delay of wide-area communication network, it can still maintain a good damping performance. The hardware controller designed can be regarded as a basic wide-area control unit embedded in the current WAMS, or can be further developed by adding wireless communication function to directly implement wide-area damping strategy.

References

1. Li Y, Liu F, Cao Y (2015) Delay-dependent wide-area damping control for stability enhancement of HVDC/AC interconnected power systems. IEEE Trans Control Eng Pract 37 (1):43–54
2. Li Y, Liu F, Rehtanz C, Luo L, Cao Y (2012) Dynamic output-feedback wide area damping control of HVDC transmission considering signal time-varying delay for stability enhancement of interconnected power systems. IEEE Trans Renew Sustain Energy Rev 16 (1):5747–5759
3. IEEE SSR Task Force (1977) First benchmark model for computer simulation for subsynchronous resonance. IEEE Trans Power Apparatus Syst PAS-96(5):1565–1571
4. Jovcic D, Pillai GN (2005) Analytical modeling of TCSC dynamics. IEEE Trans Power Delivery 20(2):1097–1104
5. Yang N, Liu Q, McCalley JD (1998) TCSC controller design for damping interarea oscillations. IEEE Trans Power Syst 13(4):1304–1310
6. Yang N, Liu Q, McCalley JD (1998) TCSC controller design for damping inter-area oscillations. IEEE Trans Power Syst 13(4):1304–1310

7. Ayres HM, Kopcak I, Castro MS, Milano F, da Costa VF (2010) A didactic procedure for designing power oscillation dampers of FACTS devices. Simul Model Pract Theory 18 (6):869–909
8. Wu M, He Y, She JH, Liu GP (2004) Delay-dependent criteria for robust stability of time-varying delay systems. Automatica 40(8):1435–1439
9. Segundo Sevilla FR, Jaimoukha I, Chaudhuri B, Korba P (2014) Passive iterative fault-tolerant control design to enhance damping of inter-area oscillations in power grids. Int J Robust Nonlinear Control 24:1304–1316
10. Gahinet P, Nemirovski A, Laub A, Chilali M (1995) LMI Control toolbox for use with Matlab. The Math Works Inc., Natick, MA
11. He Y, Wu M, Liu G-P, She J-H (2008) Output feedback stabilization for discrete-time system with a time-varying delay. IEEE Trans Autom Control 53(10):2372–2377
12. El Ghaoui L, Oustry F, AitRami M (1997) A cone complementarity linearization algorithm for static output-feedback and related problems. IEEE Trans Autom Control 42(8):1171–1176
13. Pei SC, Hsu HJ (2008) Fractional bilinear transform for analog-to-digital conversion. IEEE Trans Sig Proc 56(5):2122–2127
14. Atmel Corporation (2009) AVR 32-bit Microcontroller AT32UC3A-Type Preliminary
15. RT-LAB v.10 (2010) Opal-RT Technologies

Chapter 12
Design and Implementation of Parallel Processing in Embedded PDC Application for FACTS Wide-Area Damping Control

In this chapter, the design and implementation of parallel processing in embedded PDC (Phasor Data Concentrator) application for monitoring and stability enhancement of interconnected power system are described and examined. First, the structure of an interconnected system equipped with PMUs (Phasor Measurement Units), SMU (Synchronized Measurement Unit) and an embedded system with PDC and FACTS wide-area damping controller applications is established. Then, the fundamental modules of the embedded system is designed, and the embedded PDC application is implemented elaborately on the evaluation kit EVK1100 from Atmel®, including communicating module, data type transforming and data processing, monitoring and protection related with PDC application, as well as control algorithm and control signal output of the embedded WADC (Wide-Area Damping Control) through extern DAC (Digital-to-Analog Converter). The main program workflow of the parallel processing in embedded PDC and POD (Power Oscillation Damping) controller applications is designed and presented. Finally, the closed-loop experiment of parallel processing in embedded PDC application for FACTS wide-area damping control is carried out based on a real-time model of the typical interconnected system equipped with a shunt-type SVC device.

12.1 System Description

In this section, the device with embedded PDC and controller application can be seen as a kind of IEDs (Intelligent Electronic Devices) applied in the electric power system. The structure of the interconnected system equipped with PMUs and the device embedded with PDC and POD controller applications is illustrated in Fig. 12.1. It is established with RT-LAB® command station, the embedded systems and PMU (N60 from GE), oscilloscope (MSO2024 from Tektronix) and PC for data recording and monitoring, as well as the interfaces that interconnect them.

With modern microprocessor and communication technology [1, 2], this embedded device is designed to receive measured data from different PMUs and

© Springer-Verlag Berlin Heidelberg 2016
Y. Li et al., *Interconnected Power Systems*, Power Systems,
DOI 10.1007/978-3-662-48627-6_12

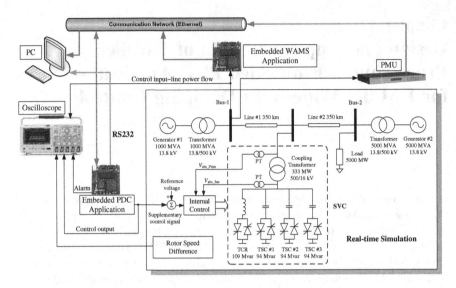

Fig. 12.1 An interconnected system equipped with PMUs, an embedded system and FACTS wide-area damping controller applications

SMUs via Ethernet communication, to handle the received data and complete control computing, and then to excite digital or analog output signals for system monitoring and wide-area control. The controller output of the embedded system is exported as an analog voltage signal, which is imported to the SVC (Static Var Compensation) device as a supplementary control signal to carry out wide-area damping performance. Moreover, the embedded system should be reprogrammable for meeting different requirements and configurations.

In Fig. 12.1, an embedded WAMS unit equipped with a boosted circuit is particularly designed to simulate the network operation of a SMU, which can sample the output voltage signal from the simulation platform with its on-chip ADC module and compute the sampled data to actual value of line power flow. This embedded WAMS unit packs the computed data in the data field of "ANALOG" in a standard C37.118 [3, 4] message and send the message in UDP packet to the embedded PDC through communication network at a reporting rate of 100 times per second.

The detailed setup of the experimental system can be seen in Fig. 12.2. During the experimentation, a PC is connected to the embedded PDC either with RS232 interface or Ethernet to receive and store the data concentrated by the embedded PDC. An analog signal proportional to one of the bus-voltage on Bus-1 is imported from the simulation platform into the PMU, which has a reporting rate at 50 times per second. The embedded PDC receives the data packets from both of the PMU and the WAMS unit and perform monitoring and control algorithm in parallel. The oscilloscope is used to measure the signals of control input, control output, rotor speed difference and the alarm level 2 generated by the embedded PDC.

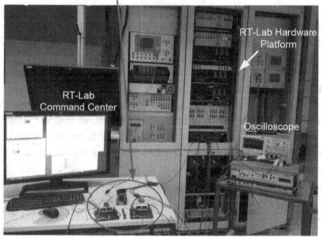

Fig. 12.2 Setup of experiment system

12.2 Design of the Embedded System

Generally, as shown in Fig. 12.3, the embedded system has the following fundamental modules:

- Data input components;
- Data storage and processing units;
- Signal or data output components;
- Tool kit for device programming and on-chip debugging.

The data buffers are high-speed registers to store data for communication. In the meanwhile, the I/O components are used for normal analog or digital signal transmitting. Program designed by developer can be written into a program memory

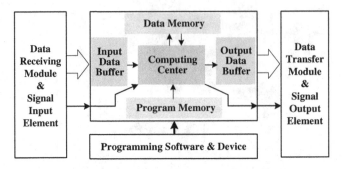

Fig. 12.3 Structure of the embedded system

with brown out protection by using programming tools. The computing center, i.e., CPU, executes the instructions in program memory, calculates the incoming data in allowable data types, sends the computing results to data memory and delivers control commands to output modules.

In the embedded PDC, the receiving data messages can be considered as input data, thus Ethernet receiving block together with register interface are the input components. The outgoing data or signal can be transmitted to external devices through SPI (Serial Peripheral Interface), GPIO (General Purpose In/Output), and RS232 (Recommend Standard 232) interface or even Ethernet transmit block. The program memory is mainly the programmable ROM or reprogrammable Flash and the data memory can be SRAM or extern storage equipment.

The PDC and FACTS controller applications are embedded in an evaluation kit named "EVK1100-AT32UC3A" of product line AVR 32-bit from Atmel Corporation. With high-performance and high-code-density 32-bit RISC microprocessor core running at frequencies up to 66 MHz, it can support real-time operating systems and achieve high computation capability by using various library instructions and functions. Besides, the AT32UC32 microcontroller incorporates on-chip memories (512 Kbytes reprogrammable Flash memory and 64 Kbytes SRAM as data memory) for secure and fast access [5].

In the evaluation kit EVK1100, the 10/100 Ethernet MAC module enables network-based applications while the CPU provides fast access to GPIOs and other standard interfaces such as SPI and RS232. Using the IEEE 1149.1 compliant JTAG interface and the assorted programming tool JTAGICE3, developer can write the program memory (521 Kbytes Flash) to program the microcontroller through the free IDE (Integrated Development Environment) of "AVR Studio® 5.1" provided by Atmel Corporation, which supports C/C++ and assembler codes with integrated compilers. And necessary information can be displayed by the 20*4 alphanumeric LCD and LEDs.

Furthermore, AVR studio 5 provides especially a collection of production-ready source codes including peripheral drivers, communication stacks and application specific libraries called "AVR Software Framework" [6], which simplified the establishing of a high-required complex project.

12.3 Implementation of the Embedded System

In this section, the software and hardware implementations of a FACTS POD controller-embedded PDC are stepwise elaborated in view of functional design and requirement.

12.3.1 Data Receiving via Communication Network

As mentioned in Sect. 12.1, the incoming data of the embedded system are messages sent from PMU and SMU via network communication. Under this technical requirement, network-oriented applications are included in the project libraries and specific functions are programmed for utilizing the Ethernet MAC block on the EVK1100 board and performing data receiving or data transmitting.

The block diagram of on-chip memory (SRAM) and MACB module are shown in Fig. 12.4. Frame data is transferred to and from the MACB through DMA (Direct Memory Access) interface, which enables efficient data transmission between a peripheral and memory by using high speed busses. All transfers are 32-bit words and may be single accesses or in bursts of four words [5]. Besides, a socket is always necessary to encapsulate the data from application layer to transport layer or extract data from transport layer to application layer.

Normal PMUs are configured to deliver all the messages to secondary devices through multicast using UDP protocol. However, a PMU has no information about the devices which it is going to communicated with, such as IP address and MAC address, at every beginning. A PMU can send C37.118 data messages to a specific device, only when it gets a standard command from the device and then finishes an ARP process successfully. Figure 12.5 represents a general communication procedure between PMU and embedded system.

Similar with the PMU, the embedded system also has no information about the PMU in its ARP cache (or ARP table) which is located in its SRAM. So the system must complete an ARP process to update its ARP cache with the IP- and MAC

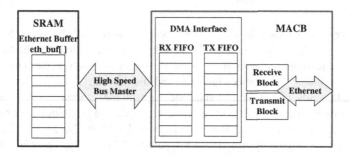

Fig. 12.4 Block diagram of Ethernet MAC block and SRAM

Fig. 12.5 Communication
procedure between PMU and
embedded system

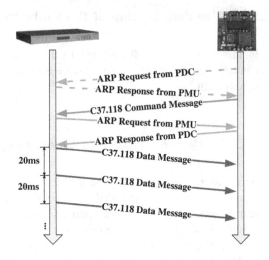

ARP Request from PDC

ARP Response from PMU

C37.118 Command Message

ARP Request from PMU

ARP Response from PDC

C37.118 Data Message

20ms

C37.118 Data Message

20ms

C37.118 Data Message

address of PMU, before it sends a standard command. The ARP process requires at least two serial work-cycles: broadcast and then receiving. A parallel working progress is not possible because of the reaction time interval of both PMU and embedded system. In order to increase the system efficiency as well as speed up the initialization, a Write-ARP-Table process is programmed to write the associated IP and MAC address of PMU directly into the ARP cache of the embedded system, when the information of PMU is already known.

The software interface is programmed in C-language on the development environment of AVR Studio® 5.1. After initializing all the required modules on the EVK1100 board, some preliminary processes are to be done to establish the connection between PMU and the embedded system. Figure 12.6 is the program workflow. These preliminary processes can be seen as the initialization of network communication. The C37.118 messages from a given PMU are identified by SYNC and IDCODE.

12.3.2 Data Processing

After a receiving work-cycle, a C37.118 message is stored in Ethernet buffer, which is defined as an U8-type data array in memory. The embedded system must have the ability to read out the desired data from the buffer and convert them to variables which are available for control algorithm and monitoring.

The data fields of a C37.118 data message in the Ethernet buffer "eth_buf" includes 3-phase measured data in rectangular format and frequency information in 16-bit signed integer form, and SOC together with FRACSEC in even 32-bit integer form, which are separately stored in U8-type-based data array. Therefore, the data bursts must be processed back to their original data types in C37.118 standard at

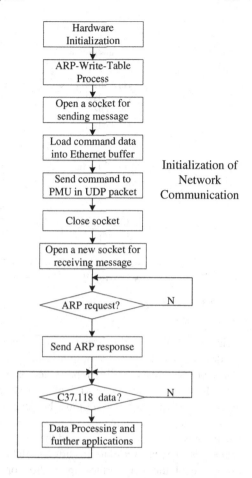

Fig. 12.6 Program workflow of a point-to-point communication procedure

first, so that the indicated value of each data can be recognized and read out by the microprocessor correctly, as shown in Fig. 12.7.

To convert the 8-bit bursts to variables with specific data types, there are two possible approaches: (1) define a pointer with desired data type and assign the address of the first burst of this data field to the pointer, then assign the target variable indirectly by making use of the pointer; (2) copy the contents of the bursts directly to target variables through the source function "memory copy" provided by Atmel®.

With the converted data in correct data type, the microprocessor can make different kinds of calculations to prepare required data for further applications. The measured data of C37.118 are represented in rectangular format with real- and imaginary values. But system monitoring requires the values of magnitudes, frequency, and phase angles. So the desired values must be processed. It is programmed to call the function "atan2" and "sqrt" and assign the return value to target

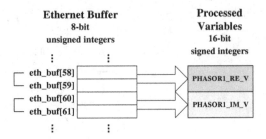

Fig. 12.7 Conversion of the 8-bit bursts to original data form

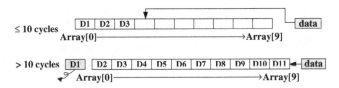

Fig. 12.8 Data storage strategy in array

variable. The library file "libm.a" must be added into the project GCC libraries by the developer so that the C-compiler can invoke the function correctly. Besides, the data type "floating point" is represented as 32-bit "double" in AVR 32-bit micro-controller. Thus, the above variables to store the calculated results such as frequency, angle and magnitude must be defined in double-type.

Since only one sequence of measured data can be obtained during every receiving work-cycle, it is designed that the embedded system must store at least 10 sequences of measured data in different data arrays (shown in Fig. 12.8), so as to make use of them in further application or send them to external memories or devices through various transmitting interfaces. Moreover, the original DSP library from Atmel® contains no function for complex analysis, so some functions are also written to carry out operations of matrix with complex operands in this paper. All the independent developed functions or files are given clear declaration in the developer announcement or commentary.

12.3.3   Monitoring and Protection

After data processing, the embedded system obtains sufficient data from a PMU to run dynamic monitoring of a fixed note in the power system. Figure 12.9 depicts the concept of a point-to-point monitoring system. In a single point monitoring work-cycle, the embedded system catches C37.118 message from a fixed PMU at first, then calculates the target quantities and analyzes or estimates the system

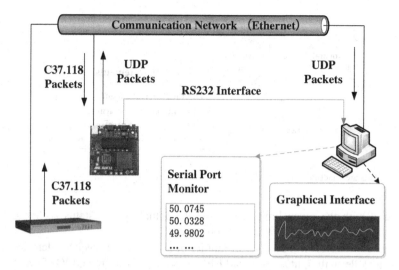

Fig. 12.9 Concept of the monitoring system

performance through the processed results, and finally sends the results to external equipment via different interfaces.

The computing results are transmitted from the EVK1100 board to a PC through either Ethernet or RS232 interface using USART (Universal Synchronous Asynchronous Receiver Transmitter) hardware, which is the most common serial interface in the computer system. The data value that is going to be displayed with serial port monitoring software or LCD must be converted into readable format, not in hexadecimal or binary code. So each symbol is decoded in ACSII-code with associated functions. Because all the results are defined in 32-bit double-type, functions to transform double-value to ASCII or to display the symbol on the LCD are written and added into original source code file. In addition, the microcontroller is connected to the alphanumeric LCD with SPI, which has a baud-rate at 12,000 Kbps. In the meanwhile, the baud-rate of USART_1 is configured at 115,200 bps.

Compared with serial interface, the transmission through communication net-work is much more efficient and advantageous according to its transmission rate up to 100 Mbps in a LAN. The microprocessor can pack the results in a normal UDP packet and deliver the packet to a PC, in which the IP and MAC address are already known, and visualize the measured data by using various graphical interfaces.

Except the FACTS devices, there are also some other power electronic devices (e.g. relays or power switches) equipped in the interconnected system to protect the local subsystem from big disturbance in the power system. The embedded system must be able to detect the problematic operation by judging measured data from PMU, generate alarm signals and give command to the power electronic devices. As shown in Fig. 12.10, the actual system status is represented with three levels: normal operation, alarm level 1 and alarm level 2. The processed quantities are

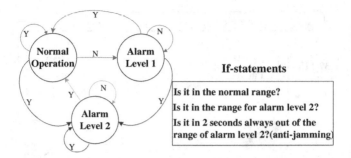

Fig. 12.10 State machine for fault detection and protection

compared with the given limit values and the actual system status is then determined though if-statements and logic operations.

In the condition that frequency instability of alarm level 1 has been detected, the microcontroller will light an on-board LED by setting one of the GPIOs from "1" to "0" (LED0_GPIO and LED1_GPIO). If the disturbance is so big that it reaches alarm level 2, another GPIO will be set to "1" (3.3 V base AVR32_PIN_PC04) to drive the protection devices and the second on-board LED is inspired. However, some disturbances can cause periodic changes on the measured frequency. So, the alarm level 2 can be deleted only when the measured frequency keeps in allowed range for 2 s.

12.3.4 Wide-Area Damping Controller

Now the power oscillation damping (POD) controller must be realized in C-program as a function which performs control algorithm. Actually, the transfer function of the POD controller designed in continuous time-domain must be converted from s-domain into the equivalent z-domain for the microcontroller programming. The bilinear transformation [7], is one of the advantageous and simplified methods that can replace the continuous time computation of integrals with numerical approximations [8]. In fact, the design of the POD controller has been mentioned in Sect. 11.3.1.

For each iteration step, the microcontroller requires at least three values: the last input value, the actual input value and the last output value. Generally, the initial output value can be regarded as "0" and each calculated result must be stored for the next iteration step. The input value of the POD controller is derived from the discrete data through data processing, which has been stored in data arrays sequentially and can be read out during each iteration step.

Because the POD controller is designed with several sequential transfer blocks, the C-program of the control algorithm is also processed with serial iteration

Fig. 12.11 Program
workflow of control algorithm

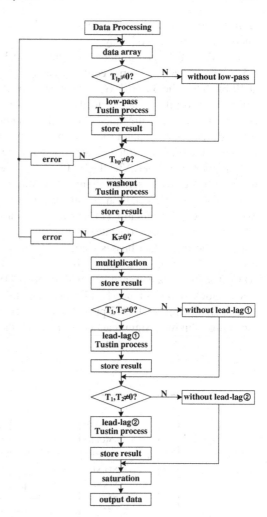

subprograms: a low-pass filter, a washout filter, a gain, two lead-lags and a satu-ration unit. The program workflow can be seen in Fig. 12.11.

12.3.5 Control Output Through SPI and External DAC

The output data U_{ALGO} of control algorithm is just a variable stored in the SRAM of the microcontroller. It cannot be read out by the external device directly and must be exported as an analog voltage signal to drive the FACTS device (SVC) equipped in the power system. However, the EVK1100 board incorporates no on-chip DAC (Digital-to-Analog Converter) and cannot generate desired analog signal through its

GPIOs, too. Therefore, an external DAC (MCP4812 from Microchip®) chip is imported to the embedded system as a solution.

The MCP4812 dual buffered voltage output DAC operates from a single 2.7 to 5.5 V supply with SPI serial communication interface [9] and has a data resolution at 10-bit. Thus the 32-bit U_{ALGO} must be converted to a 16-bit digital value D_{ALGO} and then transmitted via SPI to the external DAC. Furthermore, U_{ALGO} can be a negative value. Since the range of output voltage U_{DAC} of MCP4812 is [0, 4.096 V] [9] and the DAC cannot process negative operands. A subtraction circuit is designed to obtain the voltage output which is equivalent with the value of U_{ALGO}.

The microcontroller AT32UC3A0512 provides a serial peripheral bus interface that supports up to 15 serial peripherals in Master or Slave mode [5]. It operates as a shift register which serially transmits data to other SPIs. During the data transfer to external DAC, the on-chip SPI acts as the "Master" which controls the data flow, while the SPI of DAC operates as the "Slave" which has data shifted into by the master. The three data busses (SPCK, MOSI and NSS) constitute the 3-wire data transfer. The slave device DAC is selected when master AT32UC3A asserts its NPCS0 signal. Data transmission through MOSI tunnel is synchronized by the SPCK clock signal. Figure 12.12 represents the application diagram of SPI block.

The DAC chip MCP4812 provides a write command that consists of 16-bit word including 4-bit most significant configuration bits and 12-bit data bits. In Fig. 12.13, the write register is rewritten with the 16-bit binary values.

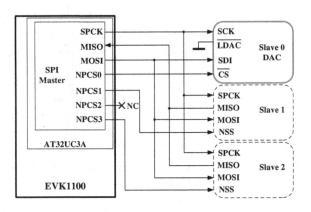

Fig. 12.12 Block diagram for SPI as master and DAC as slave

Bit 15														Bit 0
$\overline{A/B}$	X	$\overline{GA}$	$\overline{SHDN}$	D9	D8	D7	D6	D5	D4	D3	D2	D1	D0	X X
config bits						10-bit data bits from D_{ALGO}								unknown
0	0	0	1	D9	D8	D7	D6	D5	D4	D3	D2	D1	D0	0 0

Fig. 12.13 Structure and setup of 16-bit write command register of MCP4812

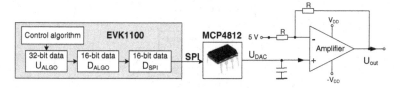

Fig. 12.14 Operation process of the DAC block

Since the output signal U_{ALGO} is limited with 0.2 V saturation, the signal detection by RT-LAB® and the accuracy of detected data could be impeded. A positive gain K_d is added for higher precision and simplification of signal detection. The real value of computing output U_{ALGO} can be converted to the digital value D_{ALGO} as

$$D_{\text{ALGO}} = 125 \times (U_{\text{ALGO}} \times K_d + 5) \tag{12.1}$$

The gain K_d is set on 15.8 in the program design in order to obtain a high data accuracy while avoiding overflow. Before the SPI data D_{SPI} is calculated and transferred to the DAC chip, the computed data D_{ALGO} must be preprocessed from 32-bit double-type to 16-bit U16-type ("unsigned short integer"), because either SPI or DAC requires operands in U16-type. The whole operational process to export the analog output voltage U_{out} is illustrated in Fig. 12.14.

12.4 Parallel Processing of the Embedded System

The applications are embedded in the EVK1100 in parallel processing and the program workflow is given in this section. The parallel processing in embedded PDC and POD controller applications are predefined in different callback functions associated with diverse sockets. A socket is a software interface between application and transport layer. One socket orients one port and is linked with one or more callbacks. A callback is actually a subroutine or a function called out by the program of socket, when the data transmission is finished. For sending a data packet without reliable transmission protocol, the callback function is usually not necessary. But in the mission, that to receive data packets and to execute corresponding application, a callback is indispensable.

Therefore, the application to be embedded into the system can be included or designed in a callback and called out during the data receiving work-cycle. The microcontroller can monitor the Ethernet messages and sort the target messages out for different sockets through the port number encapsulated in the header segment. It is defined in the main program that the monitor and protection applications of PDC are designed in socket 1 using local port 4714, and POD controller application is

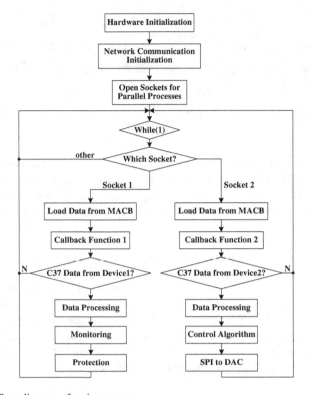

Fig. 12.15 Flow diagram of main program

related to socket 2 using local port 27895. The flow diagram is illustrated in Fig. 12.15.

The DMA controller-based MACB enables a data transfer without CPU intervention, thus the MACB block can receive data packet independently from CPU while the microcontroller is running callback functions (shown in Fig. 12.16). This kind of parallel processing can reduce the packet-loss ratio during the transmission and enhance the system working efficiency.

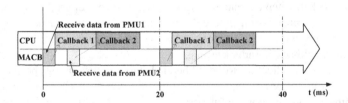

Fig. 12.16 Parallel processing architecture of the embedded system

12.5 Experimental Result

With the real-time platform RT-LAB® [10], which is fully integrated with MATLAB/Simulink®, the embedded system can be conducted into real-time HIL (Hardware-In-the-Loop). In this section, the control performance and the PDC application of the parallel processed embedded system are experimented under a big disturbance (line-to-ground fault) in the real-time test environment, and the experimental results will be presented in detail.

Figure 12.17 represents the dynamic responses of the bus-voltage magnitude, which are measured with PMU and collected and converted from rectangular format to polar format with the embedded PDC application. Apparently, the magnitude deviations of the bus-voltage caused by the inter-area oscillation continue to appear in the measurement result in the open-loop simulation. When the embedded POD

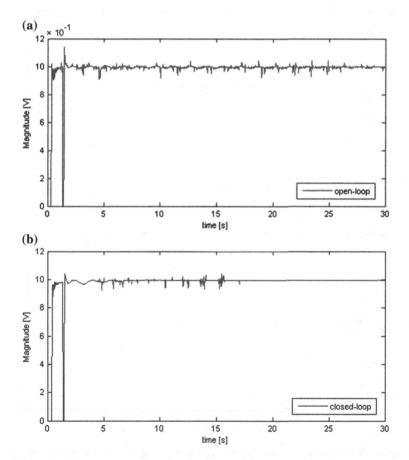

Fig. 12.17 Waveform of the measured bus-voltage magnitude in the interconnected system, **a** without embedded POD controller application; **b** with embedded POD controller application

controller is employed in the real-time simulation, the disturbance is obviously reduced and the inherent dominant inter-area mode can return to the acceptable level of the operation security. The damping effect of the controller can be seen from the waveform in Fig. 12.17b.

Figure 12.18 shows the waveform of the bus-voltage frequency measured with PMU and computed with embedded PDC. Generally, the normal frequency of the bus-voltage is 60 Hz. The initial value at 65.5 Hz is the environment noise of the RT-LAB hardware platform®, which exists inevitably before and after the simulation period. From Fig. 12.18a, it can be found that the inter-area oscillation leads to strong frequency oscillation during the open-loop simulation. However, the experimental result in the closed-loop real-time simulation reveals the damping performance of the embedded POD controller application clearly. The disturbance of the frequency is eliminated within 8 s. This verifies the damping performance of

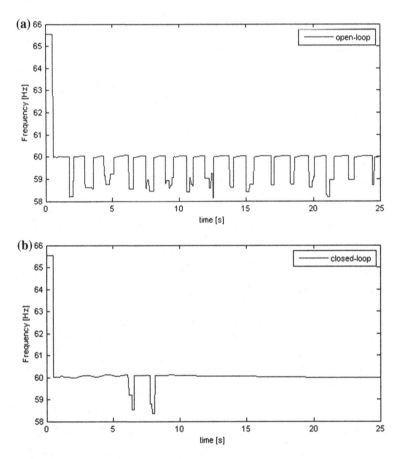

Fig. 12.18 Waveform of the measured bus-voltage frequency in the interconnected system, **a** without embedded POD controller application; **b** with embedded POD controller application

the embedded POD controller application and the embedded PDC application which is processed in parallel.

Figure 12.19 illustrates the waveform of the measured phase angles during the open-loop and closed-loop real-time simulation. Since the nominal frequency of the PMU is configured at 50 Hz, for a normal bus-voltage with 60 Hz standard frequency, the curve of the measured data is a triangular wave within [−180°, +180°] which has a period of 0.1 s. In Fig. 12.19a, the disordered waveform indicates that the periodic progress of the measured phase angle is disturbed, which means the system stability is strongly damaged by the inter-area oscillation caused by the line-to-ground fault. But Fig. 12.19b shows a different behavior of the measured data during the closed-loop simulation. By inserting the embedded POD controller into the power system, the negative impact of the inter-area oscillation on the bus-voltage is subdued. The variation of phase angle returns to a periodic procedure and the system is back to acceptable operation again.

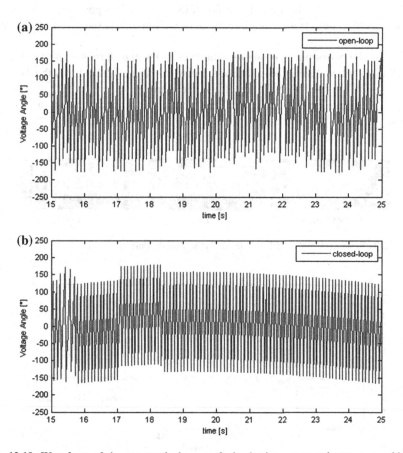

Fig. 12.19 Waveform of the measured phase angle in the interconnected system, **a** without embedded POD controller application; **b** with embedded POD controller application

Table 12.1 Frequency range of different system statuses

Normal operation	Alarm level 1 (general)	Alarm level 2 (protection)
59.05 Hz < f<60.05 Hz	f ≥ 60.05 Hz or f ≤ 59.05 Hz	f ≥ 60.1 Hz or f ≤ 59.1 Hz

By monitoring the time-dependent waveform of the measured magnitude, frequency and phase angle, it is designed to employ the frequency as the reference quantity in order to determine the system status. Table 12.1 lists the information of the frequency range in the software design of the protection application in the embedded PDC.

The alarm indication LEDs in Fig. 12.20 reveal the monitoring and protection performance of the embedded PDC, which are introduced in Sect. 12.4. When the observed frequency of the bus-voltage remains in the normal level, none of the LEDs is on. When the system experiences a small disturbance, but the frequency is still acceptable, only the on-chip LED1 shows an alarm signal. As soon as the embedded PDC detects a serious excursion of the frequency, the second LED turns on.

Except the performance of alarm indication LEDs, the alarm signal of the GPIO pin is also verified during the real-time simulation. Figure 12.21 shows the dynamic responses of the control input signal and the alarm signal of AVR32_PIN_PC04 during an open-loop real-time simulation, which experiences a big disturbance at 5 s.

Normal Operation   Alarm Level 1   Alarm Level2

Fig. 12.20 Responses of the LEDs under different system statuses

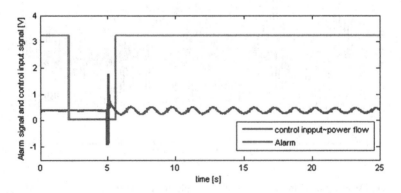

Fig. 12.21 Alarm signal and control input signal in open-loop real-time simulation

Because of the initial value of bus-voltage frequency (65.5 Hz) at the beginning of the experimentation, the initial status of the system that detected by the embedded PDC is alarm level 2, shown in Fig. 12.21. As mentioned in Sect. 4, the AVR_PIN_PC04 signal has an anti-jamming time interval of 2 s. Therefore, the alarm signal is set low until 2 s and remains low when no power oscillation is detected. Obviously, the embedded PDC indicates alarm level 2 at 5.5 s, i.e., about 0.5 s time-delay after the line-to-ground fault. This time-delay is related to the measurement accuracy of PMU and the data transmission time via communication network.

Figure 12.22 represents the dynamic responses of the alarm signal and the control input signal when the system with embedded POD controller experiences a line-to-ground fault at 1 s. It can be found that the embedded PDC application cannot detect the system instability correctly in the first 4 s. Because the embedded POD controller reduces the excursion of the bus-voltage frequency and the measured data from PMU cannot reach the range of alarm level 2, which can be seen in Fig. 12.22b. However, the embedded PDC application can recognize the oscillation after the first 4 s and clear the alarm when the system returns to the allowable operation. This verifies the performances of the embedded parallel processes again.

Furthermore, computing time and work-cycles of all the processes and subprograms are tested and listed in Table 12.2. It can be found that a 32-bit microcontroller needs at least 1 ms (millisecond) to carry out one working period of the parallel processing, when the embedded system receives two data packets from the embedded WAMS unit and one data packet from PMU within 20 ms. Therefore, the embedded system may support up to 15 PMUs while running the same wide-area damping performance in parallel. Moreover, it can be found from Table 12.2, that it takes a relative long time by using the RS232 interface to monitor the system. Thus the delay of the alarm signal will be increased when the RS232 transmission is chosen for system monitoring.

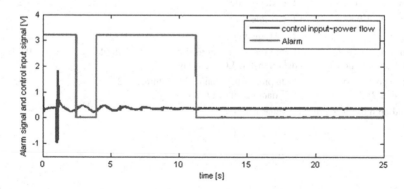

Fig. 12.22 Alarm signal and control input signal in closed-loop real-time simulation

Table 12.2 Test results of the work-cycle and computing time

Function	Tasks	Cycles	Time(ms)
Hardware initialization	PM, USART, LCD Display, MACB, SPI	120,666,654	1828.28283
ARP-write-table	1 pair of IP and MAC	2473	0.03747
Mag. & Ang. & Freq. Calculation (together)	3 × Voltage	9244	0.14006
	3 × Current		
	1 × Frequency		
Data storage	10 × 6 × Magnitude	5108	0.07739
	10 × 6 × Angle		
	10 × 1 × Frequency		
Conversion of data types	3 × Voltage	1314	0.01991
	3 × Current		
	1 × Frequency		
Control algorithm	POD processing	4072	0.06170
SPI output	16-bit data conversion and data transmission	448	0.00679
LCD display[a]	Number of packets	16,611[a]	0.25168
Calculation of angle (individual)	atan2(), transformation from radian to angle	2848	0.04315
Protection	Alarm level 2	508	0.00770
Protection	Alarm level 1	632	0.00958
Protection	Normal operation	757	0.01470
Complex multiplication	Matrix (6 × 6), Vector (6 × 1)	24,838	0.37633
Complex multiplication	Matrix (6 × 6), Matrix (6 × 6)	133,959	2.02968
Matrix transpose	Matrix (6 × 6)	2844	0.04309
Monitoring through UDP transmission	Send a 64-bit double-value to PC monitor	3356	0.05085
Monitoring through RS232 transmission	1 × Magnitude	281,562[a]	4.26609
	1 × Angle		
	1 × Frequency		
udp_monitor_pmu (callback function 1)	Data processing, protection, UDP data transfer, LCD display[a]	19,956[a]	0.30224
udp_algo (callback function 2)	Data processing, control algorithm, SPI data transfer, LCD display[a]	23,379[a]	0.35423

[a]could be more if the display value get greater

12.6 Summary

In this chapter, the software and hardware design of the parallel processing is implemented in an evaluation kit of EVK1100. The structure of the interconnected system is introduced first. Then the design and implementation of the embedded system is stepwise elaborated in view of functional design and requirement. The functional modules of the PDC application are programmed and implemented, which includes communication, data processing, monitoring, and protection. Afterwards, the control algorithm of the designed POD controller is developed referring to the Tustin method. The data processing for SPI data transmission and the external DAC circuit are designed to export the analog control signal. Finally, the parallel processing of both of the PDC and POD controller application is integrated in the program of the embedded system.

Real-time experiments are performed to verify the parallel processes of FACTS wide-area damping controller and data concentrator implemented on an evaluation kit. This verifies not only the damping performance of the embedded POD controller application, but also the embedded PDC application which is processed in parallel. The FACTS controller as well as the PDC application of the embedded system has been developed by adding wireless communication function to directly implement wide-area damping strategy. Moreover, improving the damping performance of the embedded FACTS POD controller by enhancing the reporting/sampling rate of the WAMS units will be taken into further consideration.

References

1. Wang S, Meng X, Chen T (2012) Wide-area control of power systems through delayed network communication. IEEE Trans Control Syst Technol 20:495–503
2. Zurawski R (2007) From wireline to wireless networks and technologies. IEEE Trans Ind Inf 3 (2):93–94
3. Das S, Sidhu TS (2014) Application of compressive sampling in synchrophasor data communication in WAMS. IEEE Trans Ind Inf 10(1):450–460
4. IEEE Standard for Synchrophasors for Power Systems (2006) IEEE Standard C37. 118-2005, March 2006
5. User Reference Manual (2011) AVR 32-bit Microcontroller AT32UC3A Preliminary. ©2011 Atmel Corporation. 32058 J-AVR32-04/11
6. Atmel Corporation. Atmel® AVR Studio® 5 Intuitive. Easy. Efficient. http://www.atmel.com/microsite/avr_studio_5/. Accessed 25 Aug 2015
7. Pei SC, Hsu HJ (2008) Fractional biliner transform for analog-to-digital conversion. IEEE Trans Signal Process 56(5):2122–2127
8. Kamaraju V, Ralasimham RL (2009) Linear systems-analysis and applications. ©2009 I. K. International Publishing House Pvt. Ltd. ISBN: 978-81-80026-71-8
9. User Reference Manual MCP4812 Datasheet. ©2011 Microchip Corporation
10. RT-LAB v.10 (2010) Opal-RT technologies

Printed in the United States
By Bookmasters